416

MASTERS'S WALLABY
HALMATURUS MASTERSI.
Queensland.

CAPTURING
NATURE

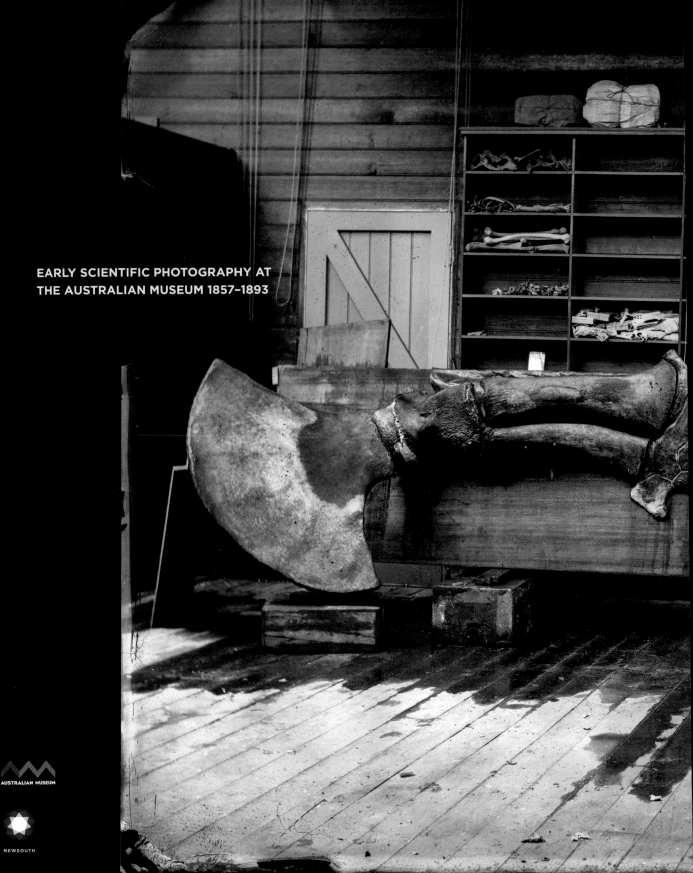

EARLY SCIENTIFIC PHOTOGRAPHY AT
THE AUSTRALIAN MUSEUM 1857–1893

AUSTRALIAN MUSEUM

NEWSOUTH

V304

CAPTURING
NATURE

Vanessa Finney

FOREWORD

Kim McKay AO
Director and CEO
Australian Museum

It seems unimaginable today that there was a time when you didn't just reach into your pocket for your smartphone and happily snap away, capturing every situation or location that interested you – complete with good auto lighting, special effects and vibrant colour (and the option of digitally recording the action)!

This book and the exhibition of the same name brilliantly reflect on the pioneering days of photography at the Australian Museum, from 1857 to 1893, at a time when new animal discoveries were being made and Australia's fauna was captivating the imagination of the science community around the world.

During this era, it took a 'team' to compose a photograph for posterity. Photographic equipment was heavy and cumbersome, images were recorded on glass plates, and darkrooms held mysterious secrets as figures and scenes were teased out and revealed by the use of chemicals. We are thankful now for the foresight of the Australian Museum's first photographers, who have shared those moments in time with us so vividly through this early black-and-white imagery.

The Australian Museum's chief archivist and librarian, Vanessa Finney, has dug deep in the museum's vaults to uncover hidden treasures recorded on these glass plates, along with the stories of the museum's curators, taxidermists and scientists who worked on the growing collection of mammals, fish, birds,

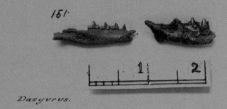

Dasyvrus.

reptiles, insects and palaeontological specimens during the last half of the 19th century.

As this book recounts: 'The first recorded mention of photography at the Australian Museum dates from 1857, when a request from a Mr Glen Wilson to photograph fossils "and some other objects" is duly noted in the Trust Minutes. By the early 1860s the museum had established its own photographic program under [Gerard] Krefft's direction, and so the next request to undertake photography was met with a defensive rebuff'.

Gerard Krefft, one of the Australian Museum's early curators, and a favourite figure in the institution's history, introduced the formal study of science to the museum. In 1856–1857 Krefft had participated in Blandowski's expedition to explore the Lower Murray and Darling Rivers, and it was during this field adventure that he was initially exposed to photographic technology. Although the results were not successful, the experience inspired in Krefft a desire to explore photography's potential as a new medium for recording the natural world.

Until this time, scientists had relied on their own drawings or those of scientific illustrators. Krefft had become a competent drawer and painter, but by the 1860s he was working with taxidermist Henry Barnes to take the museum's first photographs.

Finney recounts that it was the series of fossil digs at Wellington Caves in central-western New South Wales in 1866, 1869 and 1881 that 'cemented the usefulness of photography for the museum, and for Krefft'. Excavations at this significant site yielded the remains of some of Australia's unique megafauna, and photographs of the specimens provided important evidence of the finds while also boosting the scientific standing of the Australian Museum (and its curator Krefft) at home and abroad.

Capturing Nature reveals the stories behind the photographic subjects as well as the challenges faced by the colony's early image-makers. Just as DNA technology today can reveal the deep history and stories of the animals and specimens held in the Australian Museum's collections, these original photographs create a clear picture of the ways natural science was studied and the prized specimens regarded.

Today, the museum continues to record its collection and displays by taking beautiful still images and footage of its activities on virtually a daily basis, sharing the images on social media and creating a visual archive for the future that details most aspects of our team's extraordinary work. It is there digitally recorded for all to see. But digital communication might yet have its limitations, as *Capturing Nature* reveals. These evocative early images from the Australian Museum's collections provide today's audience with an entirely new way of seeing and learning.

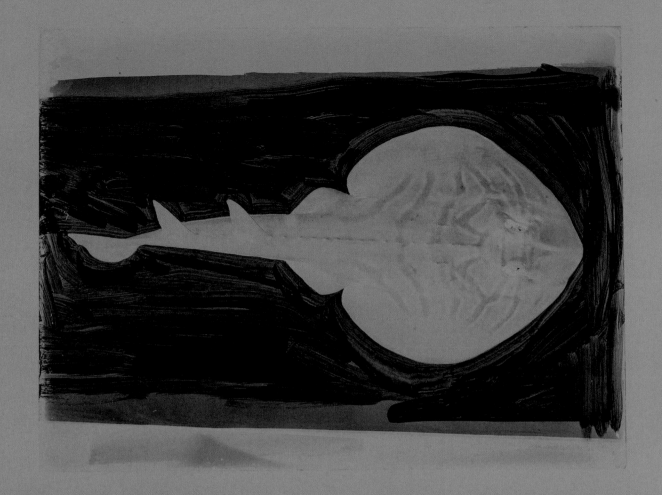

CONTENTS

FOREWORD IV

INTRODUCTION: THE MUSEUM AND THE MECHANICAL EYE 1

PART I: TIME, PLACE AND PEOPLE

CHAPTER ONE: NEW VISIONS OF THE NATURAL WORLD 11

CHAPTER TWO: THE MEN BEHIND THE IMAGES 43

CHAPTER THREE: MAKING AND MANAGING THE COLLECTIONS 81

PART II: ARTISANS AND TECHNICIANS

CHAPTER FOUR: THE AUSTRALIAN MUSEUM PHOTOGRAPHER, 1857–1893 121

CHAPTER FIVE: THE ART OF TAXIDERMY AND ARTICULATION 151

ACKNOWLEDGMENTS 186

AUTHOR BIOGRAPHY 187

LIST OF ILLUSTRATIONS 188

NOTES 190

INDEX 195

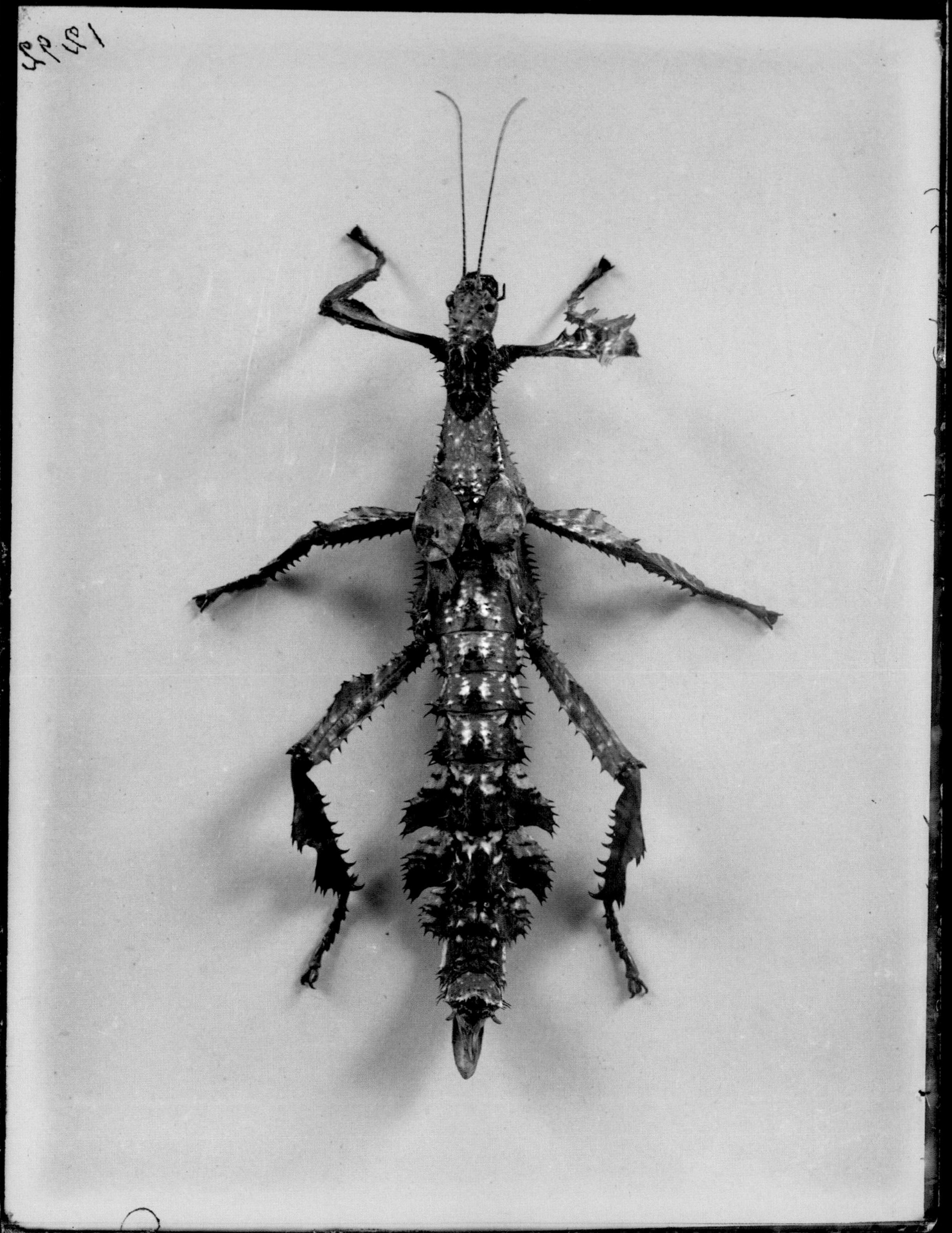

THE MUSEUM AND
THE MECHANICAL EYE

A photographic establishment is one of the
most essential parts of a modern museum.

Gerard Krefft, 1869 [1]

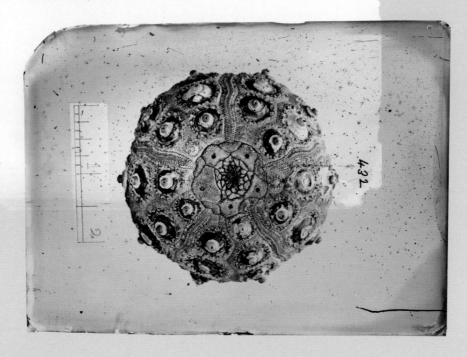

The invention of photography in the 19th century was an international sensation. Announced publicly in 1839, the new technology had the fantastic ability to capture light on glass and to provide, for the first time, an unmediated, mechanical representation of nature and the world, just as it was. The possibilities – for memorialisation, for documentation, for art and for science – were endless, and photography's cultural resonance was enormous. It quickly became a mass medium, and the promise of new optical realism changed the way Victorians saw the world. Within a decade, photography had reached Australia and become part of the way colonial Australians imagined, portrayed and viewed themselves, their homes, their landscape and their lives.

Capturing Nature is inspired by a unique record of early Australian science: the animal photography collection created at the Australian Museum in Sydney. The images range from the initial tentative experiments in the early 1860s to the time when photography was becoming an indispensable part of museum practice in the early 1890s. Beautiful, haunting and sometimes strange, this unique collection is little known outside the museum and has never before been revealed to the public.

Mostly, the photographs document the rapid expansion of the museum's specimen collections in the 19th century. They are a museum 'rogues' gallery': dozens of animals captured, mugshot style, against a white-sheet backdrop. The photos were taken in and around the museum, mostly in the courtyards and gardens to best exploit the precious light required by the photographers' rudimentary cameras. Processes were complicated and error prone; chemical recipes hard to replicate; papers brittle and unreliable; and the sunlight that was so essential to the photographs difficult to regulate and control. Images printed on paper were sometimes fleeting; chemically unstable, many quickly faded. But the negatives the men created on cut sheets of handmade glass are robust, sharp and enduring. Produced for the practical purpose of documenting newly arrived and recently taxidermied and articulated specimens, the stark, simple, portrait-style photographs contain rich incidental details of time, place and people. Although for contemporary audiences they might seem to stand alone as enigmatic artworks, each portrait tells a story and reveals its maker's mark – sometimes literally, as a thumbprint on the glass plate, accidentally pressed into the chemical emulsion that coated the glass during the development process.

Imagine the challenges faced by these early photographers. Beyond the bloody, painstaking process of preparing dead animals and bones for display in the hot Sydney sun with buckets, tubs and a basic set of tools – hammers, chisels, files and drills – they faced the added technical difficulties of capturing their images one by one using the slow, cumbersome processes of wet plate photography (see chapter four). Looking through the still-sharp and clear detail of the glass plates, it's easy to imagine that every part of the process required not only skill and expertise, but also hard work, ingenuity, coordination and team effort. Museum photography was an activity that required many participants, as revealed by the shadowy figures who sometimes appear as a blurred, ghostly presence in the photographs. The charm of the photos is in their human scale; these images acknowledge the human in scientific photography, rather than concealing it.

Page viii: Female Giant Prickly Stick Insect, *Extatosoma tiaratum*, possibly acquired in 1891. Stick insect specimens dry out and fall apart over time. As a result, the oldest stick insect in the museum's entomology collections dates back only to 1923 and this specimen no longer exists.
Photographer: Henry Barnes

Page 1: The spines on this Slate Pencil Urchin, *Phyllacanthus parvispinus*, have been removed to more clearly show details of the other parts of the animal. The photograph was taken in 1880 and published in Edward Ramsay's *Catalogue of Echinodermata in the Australian Museum* in 1885.
Photographer: Henry Barnes

Above: The long exposures needed for photographs meant that mistakes sometimes happened. The sheet behind the dolphin has moved with the breeze and you can just see the boots of two ghostly humans at each end of the table.
Photographer: Henry Barnes

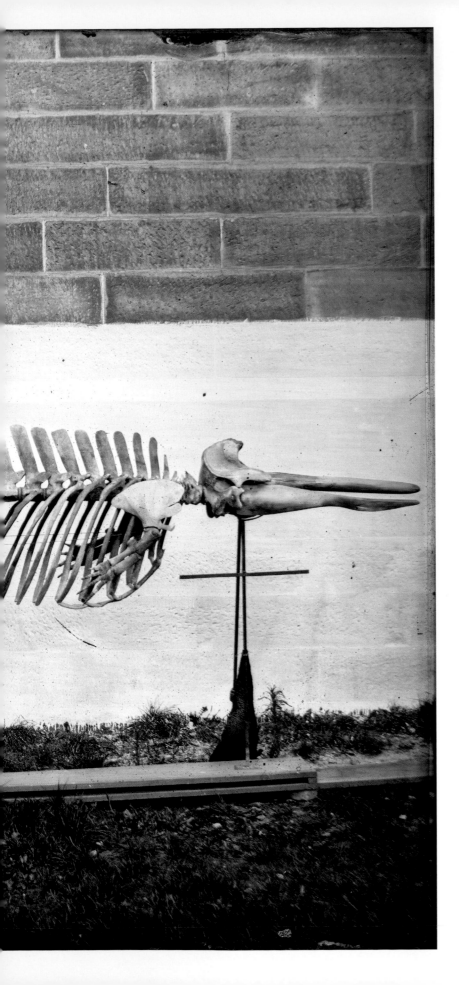

Left: In a rare moment of
whimsy, Gerard Krefft leans out
of a gallery window above the
newly prepared skeleton of a
Beaked Whale, *Mesoplodon* sp.
Photographer: Henry Barnes

Dendrolagus dorianus. Ram:
(enlarged)

For the Australian Museum, the 1860s marked a lucky confluence of skills, experience, need and technology.

In 1864 – after several years as assistant and then acting curator – Gerard Krefft was officially appointed as the museum's curator, a position he held for ten years. Krefft had encountered photography as a member of the 1856–1857 Blandowski expedition through Victoria along the Lower Murray and Darling Rivers – and possibly at the newly established National Museum in Melbourne, where he had worked as a collector and cataloguer. His new position at the Australian Museum allowed him the time to fully explore the medium and its possibilities.

In this, Krefft had the practical support of museum assistant Henry Barnes, who was able to turn his hand and eye as easily and confidently to the intricacies of taxidermy and articulation (the process of cleaning and assembling an animal skeleton) as to the complicated new techniques of photography. Both men were unafraid of failure, and they took to photography as they had to the other skills they needed to become competent colonial field and museum naturalists – developing expertise through experimentation, trial and error, and learning by doing. The thousands of plates they produced over the fifteen years they worked together represent not just an index of the museum's specimens, but also a new visual language for natural history and for the museum.

If the project was directed by Krefft, it was largely stage-managed and produced by Henry Barnes. It is Barnes who is usually credited as photographer in the associated photographic registers, and it was Barnes' role to provide the finished animal specimens.

The story of the images' creation represents a wonderfully expressive moment in the formation of Australian scientific identity. Captured and recounted visually, it is also told through the experiences of these two men as together they experimented with the new science of photography, identifying and refining its ongoing purpose at the Australian Museum. The introduction of photography at

Above: Skull of Doria's Tree-kangaroo, *Dendrolagus dorianus*, first described by Edward Ramsay in 1883. Photographer: Henry Barnes

the museum in the early 1860s was the beginning of an experiment that would change the institution.

Most of the images are plain, tightly focused portraits, high contrast and black and white. The photographs are still, stark and technocratic – rather than luxurious or sensuous in the way of earlier generations of the hand-drawn and richly coloured scientific illustrations that were made and used by scientists and natural history artists, and published in journals and books. Useful within the museum for the documentation of the ephemeral detail, texture and shape of fresh specimens, the photographic images were also important for collection management, control and comparison. Accompanying the well-used paper trail of Australian science, this collection is one of its first well-formed photographic footprints. Outside the museum, the images were a shorthand way to (relatively) quickly and widely communicate the institution's scientific status and progress. They also served to announce Krefft's own successes to his networks of mentors, collaborators and peers in museums across the European scientific world, and they constituted an important part of his self-expression as a man of science.

Krefft's successor at the Australian Museum, curator Edward Ramsay, had his own photographic project. It was less personal than Krefft's, although he also worked closely with Henry Barnes, this time deploying photography as a tool for museum management: a visual index to specimens and a record of museum work.

The collection and the story of early photography at the museum are haunted by Gerard Krefft's untimely fate, a story with a climax that unfolds partly through his photographic work. He lost the job he loved after only ten years, dismissed when his scientific ambition and poor personal skills ran up against the distrust and ultimate betrayal of his colleagues and the will and opportunism of the museum's trustees. Krefft's unfair treatment at the museum, and his personal tragedy of stubborn misunderstanding, cast a shadow over the energy and experiment of his photographic collaboration with Henry Barnes and make these images especially poignant.

The collection as a whole is about the realisation of new possibilities for museum documentation, sharing and communication – enabled by the creation, reproduction, use and distribution of information and iconography through photographic images. And it is about the content of those images, representing some of the untold stories from the museum's early productive years of systematic collection, display and public access to its natural history and cultural holdings.

Like the preparation techniques of taxidermy and articulation that shape, groom and present photo-ready specimens, photography remains an important part of the methodology and practice of natural history museums. However, curators and historians in Australia and elsewhere are only just beginning the study of photographic collections within museums and the ways they have been made, reproduced and exhibited. *Capturing Nature* takes an even rarer approach, by examining a museum through its own photographic processes as well as through its historical photography collection – seen as aesthetic objects, as documents and as museum history.

Pages 8–9: White Shark, *Carcharodon carcharias*, photographed in the museum's taxidermy shed. Photographer: Henry Barnes

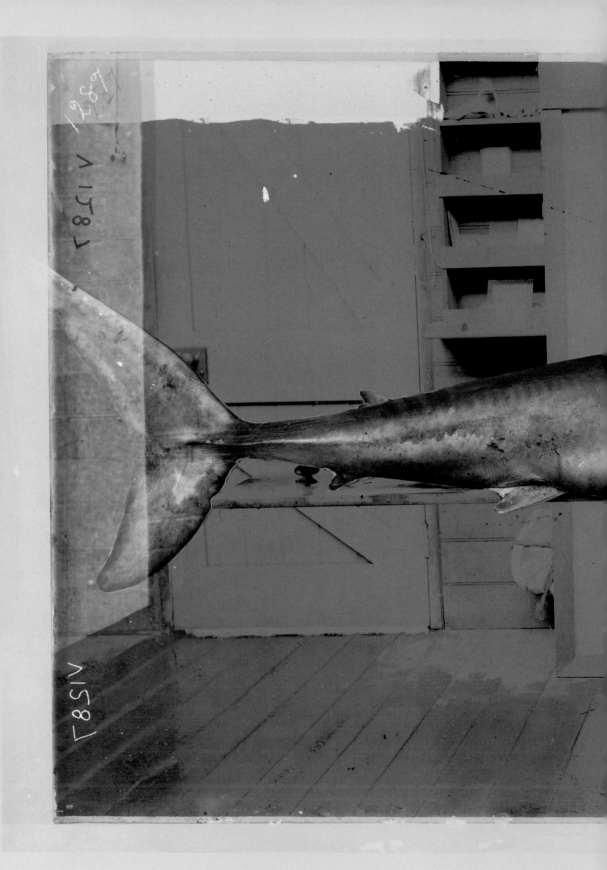

TIME, PLACE
AND PEOPLE

PART I

CHAPTER ONE

NEW VISIONS OF THE NATURAL WORLD

A HINT. – We should be glad to perceive, amongst some of the intelligent and public-spirited Colonists, more of a desire to prosecute the public weal than at present exists. Amongst other improvements, in these times, would there be any harm in suggesting the idea of founding an AUSTRALIAN MUSEUM?

The earlier such an Institution is formed, the better it will be for posterity.

Anonymous author, 1827[1]

A MUSEUM FOR EVERYONE AND EVERYTHING

In 1826, Alexander Macleay arrived in Sydney to take up the post of Colonial Secretary, bringing with him England's finest and largest private insect collection. It is no coincidence that the Colonial Museum (which was renamed the 'Australian Museum' in 1836) was founded just one year later, the first public museum in the colony. Its original inauspicious, cramped premises were two sheds at the back of the old post office in Sydney's Rocks area – although the museum and its collections would frequently have to move locations. For the next few decades, the museum was open on demand, looked after by a series of hardworking, resourceful men who collected specimens, displayed them in cases, welcomed visitors, and managed the packing and despatch of natural history items.

In the beginning, the museum acted mostly as a place where Sydney's colonial gentlemen with scientific interests could meet, and as a clearing house for 'rare, valuable and beautiful' specimens of natural history.[2] Thousands of the best specimens were sent to Europe, destined for authoritative description and classification, and then for display in English and European institutions and private collections. The museum's permanent collection was small; only a few hundred objects in a mix of natural history and curiosities.[3]

But the 1850s and 1860s saw change. The colony was newly prosperous thanks to the gold rushes that had begun in New South Wales and Victoria in 1851, sparking an influx of both money and new immigrants. Convict transportation to New South Wales had ended in 1840, and the city of Sydney was growing in size and in confidence: trade was expanding, new industries were being established, and agriculture was developing as more land was surveyed and opened for settlement. Elected members of parliament were turning their attention to local laws and conditions, and the mid-19th century ushered in the construction of government offices and civic buildings such as the Sydney Town Hall, as well as the establishment of Australia's first university (the University of Sydney). The embedding of cultural infrastructure and institution-building began in earnest. A permanent home for the Australian Museum was one such project, and in 1845 construction money was approved by the Legislative Council. As an enthusiastic museum supporter wrote in a local newspaper:

> A people struggling at the bottom of the hill, so to speak, have neither time nor means to devote to other matters than those which will serve them on the journey upward. When they achieve a footing, however,

Page 10: Banded Wobbegong,
Orectolobus ornatus.
Photographer: Henry Barnes

Page 11: Snapper, *Chrysophrys
auratus.*
Photographer: Henry Barnes

Above: The Long Gallery, around
1887. Some staff members and
their families lived on the museum
premises, and you can see a blur
that might be two children running
through the gallery (back left).
The last staff moved out of the
museum in 1888.
Photographer: Henry Barnes

and get breathing time, they bethink
themselves of other affairs than mere
wealth-collecting, and having some
leisure, turn to the best and purest
sources of human enjoyment, those
of the intellect. Then spring into
existence mechanics' institutes,
libraries, theatres, art galleries
and museums of science.[4]

The museum's staffing and funding
were boosted through the passing of
the Australian Museum Act in 1853,
which formalised museum governance
by a board of 12 elected trustees
and guaranteed funding of £1000
per annum. The institution's new
mission was encyclopedic: it was to
hold a representative collection of
Australian and foreign animal specimens
for the purposes of both comparison
and classification.

Building began on land that had once
been a convict food garden, opposite
Sydney's Hyde Park, and the museum's
first purpose-built gallery (known as
the Long Gallery) was finally opened
to the Sydney public in May 1857.

Left: Gerard Krefft (back left) standing in the Long Gallery, around 1860. In the foreground is the skeleton of the museum's first Sperm Whale specimen, acquired in 1849. You can see the elaborate wire frame attached to the floor that supported the 11 metre long whale.
Photographer: Henry Barnes

The two-level gallery was still not complete (as there were not enough funds for a staircase to reach the gallery level), but the ground floor was lined with glass-fronted cases, each one full of stuffed animals, skeletons, natural history specimens, 'relics' and 'curiosities' (cultural objects from Australia and the Pacific). The displays were confusing, and it was difficult to see inside the dark cases, but in the opening week an astonishing 10 000 people (at a time when Sydney's total population was just 40 000) came to view the crowded collections in the dimly lit space.

In addition to new storage and meeting areas, staff accommodation and secure government funding, the institution now had authority, visibility and prestige in the booming colony of New South Wales.

In the 1850s, the museum was run by a series of curators and secretaries: William Sheridan Wall (serving from 1844 to 1858), George French Angas (1853–1860) and Simon Pittard (1860–1861). Pittard served only one year, dying of consumption in 1861. Gerard Krefft then became acting curator, and was officially appointed curator in 1864. However, it was the museum's trustees who were really in charge. The first 12 elected members of the Board of Trustees of the Australian Museum were a who's who of Sydney's 'gentlemen scientists', and included geologist and clergyman William Branwhite Clarke, naturalist and medical doctor George Bennett, botanist and pastoralist William Macarthur, zoologist George Macleay and his brother William Sharp Maclcay (an entomologist, and the major architect of the *Australian Museum Act, 1853*), as well as University of Sydney chemist Professor

John Smith. The museum, which also had the active support of the governor of New South Wales, Sir William Denison, now became part of the tangled, overlapping networks of privileged and influential men who ran Sydney's politics, religion, public health, education and business – and who also oversaw its new public scientific institutions, including the Australian Museum and the Royal Botanic Garden (founded in 1816). Governor Denison was soon lobbying for extra space for the museum's growing collections: construction of a new building facing College Street began in 1861, and was completed in 1867. The extra wing tripled the museum's exhibition space, bringing with it light-filled, expansive galleries and the chance for Krefft to reorganise the collections into more systematic order. The Sydney public embraced the museum, and over the next two decades it became a pillar of the city's scientific life.

Beyond Sydney, in the 1860s and 1870s there was an Australia-wide boom in museum construction, so that by 1879 every state bar Western Australian had its own natural history and ethnography museum. Following Sydney's lead, museums opened in Victoria in 1858; in Tasmania and South Australia in 1861; and in Queensland in 1879. Western Australia's museum building came later, in 1897. As they do today, these institutions served multiple purposes, offering entertainment and education for the public, while also acting as central storehouses of specimens for serious scientific study. Their blend of science, education, pleasure and prestige helped forge and reinforce new perceptions of the natural world for scientists and the public alike.

Left: Fossil display in the entrance to the College Street wing, 1887. Photographer: Henry Barnes

Above: The museum's imposing new
College Street wing in 1887. In 1890,
the original museum building (left,
behind the trees) had a third storey
added to bring it up to the same
height as the new wing.
Photographer: Henry Barnes

Top row, left to right: Stars-and-stripes Puffer, *Arothron hispidus*; Smalltooth Flounder, *Pseudorhombus jenynsii*; Pearl Perch, *Glaucosoma scapulare*.
Middle row, left to right: Pennant Fish, *Alectis ciliaris*; Banded Seaperch, *Hypoplectrodes nigroruber*; Tripletail, *Lobotes surinamensis*.
Bottom row, left to right: Eastern Talma, *Chelmonops truncatus*; Australian Pineapplefish, *Cleidopus gloriamaris*; Striate Anglerfish *Antennarius striatus*.
Photographer: Henry Barnes

A SHARED HISTORICAL MOMENT: MUSEUMS, NATURAL SCIENCE AND PHOTOGRAPHY

Science, too, was changing. The second half of the 19th century saw the rise of scientific disciplines and the establishment of new scientific ways of seeing (with the aid of new precision instruments) and working (in universities, scientific institutions and laboratories). In the process, older, amateur versions of 'natural history' were set apart from 'natural science', which became increasingly technical and professionalised.

FROM HOBBY TO DISCIPLINE

Natural history, or the scientific study of the natural world, is at the same time a collection, a practice, and a way of ordering and seeing the world. It became popular in the West in the 16th century, when the European aristocracy accumulated personal cabinets of both natural and 'man-made' curiosities.

By the mid-19th century, its actors included amateurs and professionals; enthusiasts (both women and men); collectors; museum staff; government policymakers and funders; traders and shopkeepers; describers and taxonomists; taxidermists and storage experts; cabinet-makers; museum architects; and showmen and entrepreneurs. Across the British Empire, natural history was more democratic and widespread than ever before, its popularity expressed in everything from the craze for making fern and seaweed collection books, to naturalists' societies, well-attended public lectures and the establishment

of aquariums and zoos. Its spread was greatly assisted by the expansion in literacy and mass publishing, with the appearance of hundreds of new natural history publications – including books, newspapers, guides and specialist scientific journals.

In its academic, institutionalised form as 'natural science' it also had a solid cultural and political presence in universities and, especially, in new public museums: the British Museum, the model for colonial offspring such as the Australian Museum, had opened in 1759 (its natural history collections were moved to a new museum – the Natural History Museum in South Kensington – in the 1880s). Indeed, museums were the public face of Victorian natural science.

Ironically, at the same time as popular and scientific interest in natural history was flourishing in the 19th century, the general public was becoming more urbanised and increasingly removed from daily contact with the natural (much less the wild) world. Just as nature study became a popular pastime and naturalists' societies began to agitate for habitat protection and the establishment of areas for outdoor recreation, natural science was moving indoors – operating in laboratories equipped with scientific instruments, and in museums filled with shelves of stuffed, pickled and preserved animal specimens arranged in serried rows.

In Australia, politicians as well as the museum's scientific staff were more and more keen to foster local efforts and build and keep natural history collections locally.

Opposite: Roundface Batfish, *Platax teira*.
Photographer: Henry Barnes

Pages 22–23:
Top row, left to right: Tallfin Flyingfish, *Cheilopogon pinnatibarbatus*; Mosaic Leatherjacket, *Eubalichthys mosaicus*; Bluethroat Wrasse *Notolabrus tetricus*.
Bottom row, left to right: Spotted Bigeye, *Priacanthus macracanthus*; likely Hawaiian Giant Herring, *Elops hawaiensis*; Stout Longtom, also known as Alligator Gar, *Tylosurus gavialoides*.
Photographer: Henry Barnes

1087

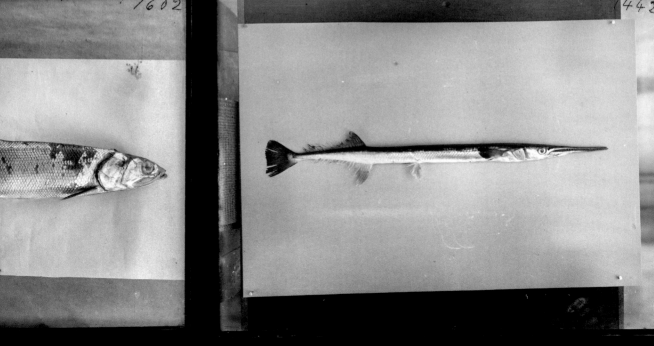

1083

1589

1602

1442

Above: Brown-eared Pheasant,
Crossoptilon mantchuricum.
This bird was presented to the
museum by Charles Moore,
director of the Royal Botanic
Garden in Sydney. For many years,
the botanic garden maintained its
own aviaries and acquired many
different birds from around the
world for display (this pheasant
is endemic to China). Those that
died came to the museum.
Photographer: Henry Barnes

Science increasingly became part of national identity. As Krefft wrote to Sir Henry Parkes (premier of New South Wales) in 1873: 'a thorough history of our Animals can only be written in this country, and in the Colony'.[5]

During the course of the 19th century, 'natural history' continued to diverge from 'natural science' through the processes of professionalisation and institutionalisation. Although individual collection and research efforts had been steadily building a substantial, shared body of knowledge about the natural world – and although museums still largely relied on amateurs for collection donations – there was now a hierarchy of scientific insiders and outsiders. In learned societies such as the Royal Society of New South Wales (founded in 1866) and the Linnean Society (founded in Sydney in 1874), and within museums, natural science was increasingly divided into discrete animal-based disciplines.

To make their mark outside the museum, scientists could best establish their scientific independence and international reputations by building their own specialist networks. They did this through participation in scientific societies, publication and the strategic exchange of both natural history specimens and letters – in which they shared scientific knowledge with their networks of collectors and museum professional peers. At the Royal Botanic Gardens in Melbourne, for example, botanist Ferdinand von Mueller's

archive holds more than 10 000 letters to and from his worldwide network of hundreds of collectors and collaborators.

DARWIN AND THE SECULARISATION OF SCIENCE

It is significant, too, that just as Gerard Krefft joined the Australian Museum, English naturalist Charles Darwin's theory of evolution arrived to shake the scientific world. In 1859, *On the Origin of Species* was published in England, marking the divergence of natural science from natural theology (that is, a religious understanding of the natural world). Within a year, British church leaders and the presidents of scientific societies were joining forces to warn the public about the moral dangers of Darwin's 'transmutation theory'. It was Darwin's belief that natural variation exists in animal populations. Due to environmental pressures, certain characteristics (or small variations) within a population are selected as favourable (since they increase the individual's ability to compete, survive and reproduce) and hence continue within a population – as it is those individuals whose offspring have a greater chance of survival. As a result, over time, the population's gene pool will change to become better suited to its environment.

Although Darwin's book arrived in Sydney as early as 1860, it took much longer for his ideas to take hold in Australia, due to opposition from leading colonial scientists such as

William Sharp Macleay in Sydney and Ferdinand von Mueller and Frederick McCoy in Melbourne. However, Krefft's experience as a palaeontologist and field collector – particularly the time he spent on megafauna excavations at Wellington Caves in New South Wales (discussed in chapter two) – meant that Darwin's ideas immediately made sense to him, and he became one of evolution's major early public supporters. Krefft corresponded with Darwin and promoted Darwin's ideas in newspaper articles and in his scientific papers. In turn, Darwin thought well of Krefft and knew his writings. The radically secular impact of Darwin's ideas would shape the history of the Australian Museum.

THE ART OF NATURAL HISTORY

Museums might be the major built expression of 19th-century natural science, but it is the animal specimens natural history museums hold that give them their object of study and mission to describe and display the natural world, item by item. However, by themselves, without labels, field notes, descriptions and catalogues, museum specimens are mute objects with little scientific meaning. The 'capture' of nature in zoological specimens involves a lot of words: precise, codified and detailed description, along with the expression and communication of that knowledge in lectures, articles, books and papers. In the 19th century, the museum-based work of naming and classification was accompanied by a growing catalogue of natural history publications – scientific texts, as well as narrative accounts of expeditions, local natural history work and field guides.

Moving beyond words, of all the sciences it is natural history that relies most on specialist visual communication for its primary tasks of comparison and classification. Natural history has always had an alignment with art and artists – well-trained amateurs as well as professional artists – and the art of natural history has been essential to the spread of its understanding of the world.

Colonial natural history illustration is a familiar part of the story of Australia's art. Collector and botanist Joseph Banks, who sailed with James Cook on the *Endeavour* in 1768–1771, equipped himself with two artists – Sydney Parkinson and Alexander Buchan – to help process and stabilise the abundance of information and specimens

the voyage was collecting. Before photography, scientists like Krefft had to be competent in drawing and drafting in order to record their experiences, catalogue specimens and publish their findings. Sydney's famous 19th-century natural history illustrators, sisters Harriet and Helena Scott, worked for both Krefft and his successor Edward Ramsay, and the second volume of their *Transformations* book on Australian butterflies and moths was published by the museum.[6]

The ability of a good scientific illustration to break a specimen into its parts still plays a vital role in scientific description. And the painstaking work of natural history illustration continues to serve a key function in the communication of natural science, and forms an important part of its knowledge base and historical record.

THE HUMAN EYE IMPROVED: SCIENTIFIC PHOTOGRAPHY

Illustrations, however, have limits. They can only record what the viewer sees. Photography, by contrast, has the potential to improve scientists' abilities to record and observe the natural world. From the 1850s, scientists were quick to recognise this potential, and the new technology emerged as an invaluable empirical tool for scientific work and measurement.

Used in this way, photography became not just a recording device but a medium of discovery. With improvements in chemical sensitivity (images were becoming ever clearer, more detailed and more stable) and camera technology, photography offered the potential to capture that which was too fast or too far away for the naked eye to see. The camera was like the human eye improved; as English scientist and photography pioneer Henry Fox Talbot described it: 'the eye of the camera would see plainly where the human eye would find nothing but darkness'.[7]

Slivers of time could be captured and studied in detail when multiple photos were taken in quick succession, as in English photographer Eadweard Muybridge's famous images of a horse jumping, part of his series *Animal Locomotion* (1887). To the naked eye, the horse's actions would have been a blur. But seen through the objective eye of the camera lens, every muscle movement and mechanic could be scrutinised in frame-by-frame detail.

At the macro scale, photography pushed the boundaries of scientific observation further into space. Gradually improving cameras, lenses and exposures meant that never-before-seen celestial objects could be captured for the first time on glass. In 1867, for example, Victorian government astonomer RLJ Ellery began to photograph the moon from Australia, and the first Australian photographs taken with a telescope recorded observations of the 1874 transit of Venus – the same celestial event that had (ostensibly) drawn Cook to the Pacific a century earlier.

By the 1880s, life could be viewed at cellular level too, with the aid of microscopes with attached cameras, a technique known as photomicrography – examples of these images appear in chapter four.

In the case of natural history, photography's first significant impact came in supplementing the sketchbook. The geological survey of Victoria in the 1850s led by Richard Daintree was the first Australian expedition to introduce photography into fieldwork studies, using wet plate equipment. The Blandowski expedition of 1856–1857 also took photographic equipment into the field, but produced few images. Because the plates had to be developed immediately upon exposure, the trays, chemicals and water baths required had to be carried into the field along with the heavy glass plates and bulky cameras. A decade later, dry plate photography (in which the glass plates could be exposed first and developed later) made field photography much easier, and the technology spread quickly to the field-based sciences, in particular anthropology.

It would be some decades before photographs could capture most animals in their natural habitats. Birds' nests, animal habitats and slow-moving or static insects could be photographed successfully, but the movement and unpredictability of most living animals made them impossible subjects for early cameras, with their slow shutter speeds and their associated unwieldy equipment.

PHOTOGRAPHY MOVES INTO THE MUSEUM

The first public museum to embrace photography as a tool for documentation and didactics was the British Museum in London. Like the Australian Museum, the initiative resulted from outside requests for photographic records of the museum's treasures. Unlike the Australian Museum, where the collection images were not on general sale, the British institution's initial photographic work was primarily a business venture. From 1853 to 1859, commercial photographer Roger Fenton worked in a glass-walled studio on the museum's roof (for access to optimal natural light) photographing thousands of the museum's iconic objects, including skeletons and zoological specimens, along with antiquities and books. At first, he worked on demand, but within a few years he also began to sell his museum images to an eager public, alongside his own commercial work. He even set up a sales booth in the museum's foyer. Fenton's images demonstrated the potential of photography to disseminate objects outside the museum, and make circulation, publicity and (potentially) research, easier.[8] As Krefft had visited staff at the British Museum in 1858, he would have known of Fenton's photographic work there.[9]

With Gerard Krefft's arrival at the Australian Museum in 1860, photography and natural history became close collaborative partners in that institution too. Over the next three decades, as the art and science of photography developed in tandem with museum changes, it became an integral part of the practice and project of museum-based natural science (as it remains today).[10]

Opposite: Skull and jawbone of the Southern Hairy-nosed Wombat, *Lasiorhinus latifrons*, late 1860s. Photographer: Henry Barnes

SNAKES ALIVE!
PHOTOGRAPHS
VERSUS
HAND-DRAWN
ILLUSTRATIONS

BLACK-HEADED
PYTHON, *ASPIDIOTES
MELANOCEPHALUS*

In 1869, Gerard Krefft published *Snakes of Australia*, which included 15 hand-coloured plates illustrated by colonial Sydney's finest natural history artists, sisters Harriet and Helena Scott, who were most famous for their exquisitely detailed illustrations of Australian butterflies and moths. Although the book was received with much acclaim, and made a vital contribution to the author's scientific reputation, Krefft was not entirely happy with the sisters' efforts, perhaps damning them with faint praise in the book's acknowledgment of 'the gifted daughters of A.W. Scott, Esq., Miss Scott and Mrs Edward Forde – [who] had done everything in their power to give correct figures of the reptiles illustrated'.[1] It is true that some of the snakes illustrated in the volume are oddly lifeless, depicted tightly coiled, reflecting the bottled and pickled specimens the sisters worked from to produce their artworks.

In marked contrast, Krefft's photograph of a live Black-headed Python, *Aspidiotes melanocephalus*, which captures in detail the natural coils of the reptile's body and the glossy sheen of its skin, is clearly the model for Harriet Scott's lithograph (above right) of that species.

Krefft's image strikingly demonstrates that photography offered the possibility of not simply replacing but enhancing the art of natural history illustration by capturing a single moment with a sense of life and movement. Drawing could still improve on the black-and-white photograph. As these images show, the artist could depict the colours of the python's skin and add extra background detail. She could even improve on nature by making the snake's head visible, when it is obscured in the photograph.

Lithography (drawing directly on stone or metal for printmaking) remained the most popular scientific illustration technique

BLACK-HEADED SNAKE,
Aspidiotes melanocephalus.

until well into the 1890s. Its precision and colour could still not be matched by photolithographs (where printing plates are made from photographs), which had begun to appear in Australia in the 1880s.

Photography and photomechanical reproduction, however, were relatively fast and cheap, and images could be mass produced – and as a result, during the late 19th century, photography gradually replaced most scientific illustration. However, for natural history field guides and taxonomic works, which rely on details of colour, size and shape that might not be clear or visible in a photograph, both coloured and line illustrations are still preferred, even today, for their superior depiction of 'averaged' detail, as opposed to the single instant that a photograph remembers.

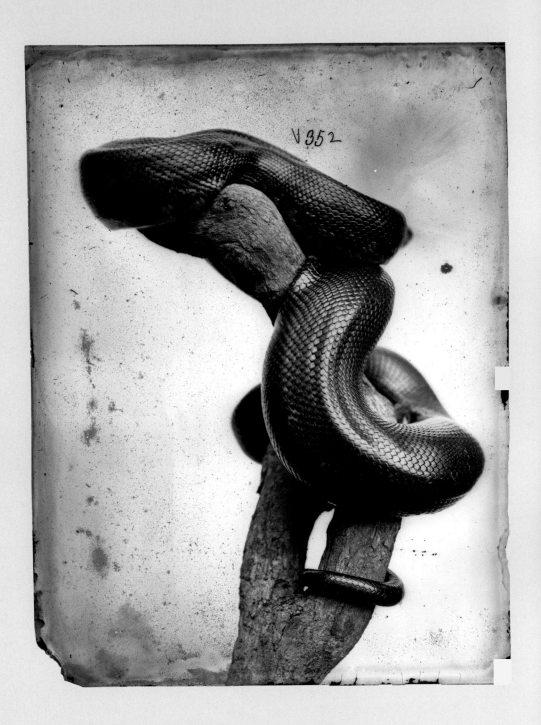

Right: Black-headed Python,
Aspidiotes melancephalus.
Photographer: Henry Barnes

Opposite: Drawing and
lithograph by Harriet Scott
for Gerard Krefft's *Snakes of
Australia*, 1869. In the process
of drawing and lithography,
the snake has been flipped.
Scott has also added a head.

PHOTOGRAPHY AS
MUSEUM PRACTICE

The first recorded mention of photography at the Australian Museum dates from 1857, when a request from a Mr Glen Wilson to photograph fossils 'and some other objects' is duly noted in the Trust Minutes. By the early 1860s the museum had established its own photographic program under Krefft's direction, and so the next request to undertake photography was met with a defensive rebuff. Photography was now part of the museum's own practice, and it would be staying under the firm control of the trustees. The unfortunate rejected applicant was William Sheridan Wall, who had been the museum's curator from 1840 to 1858 and now ran a taxidermy business in Sydney. It is likely he wanted the photographs as reference to ensure more accurate poses for his commercial natural history specimens.

> The Curator was directed to inform Mr Wall that for some time the Trustees have employed a person to take photographs of bones and other specimens of Natural History and that they would have no objection to letting Mr Wall have copies on application to the Board of Trustees.

The trustees' attempts to rein in rogue photography at the museum are a running refrain through the Trust Minutes for the next two decades. They wanted to know who was taking photographs and how they were being stored, reproduced and used. And they wanted to see and hold – and contain, control and potentially commercialise – the results. In 1878, it was again resolved 'that all photographs taken by officers of the museum shall be submitted by the Board at their first meeting; and that each member of the Board shall be entitled to obtain a copy'.[12] It is clear from their anxious deliberations over the use and distribution of photographs that the museum's trustees misunderstood photography's disruptive potential. Although they debated whether to sell photographs to the public, and approved occasional exchanges with other institutions, they seem always to have regarded photography as a transparent tool for documentation and missed its greater potential for reproducibility, commercialisation and (especially) mobility. Photographs are not just images, but objects that can be traded and exchanged, and that can travel around the world. Despite their efforts, the trustees never had the control of museum images they wished for.

ENTER GERARD KREFFT,
WITH CAMERA

When Gerard Krefft joined the museum as assistant curator in 1860, he was an experienced field collector, artist and cataloguer. (Krefft's career, together with that of Henry Barnes and Edward Ramsay, is discussed in detail in chapter two.) Significantly, he had already gained first-hand exposure to the new technology of photography on William Blandowski's expedition to the Lower Murray and Darling Rivers in 1856–1857, one of the earliest Australian expeditions

Above: *Diprotodon* molars.
Photographer: Henry Barnes

to attempt photographic documentation – largely unsuccessfully. But Krefft could see the medium's potential, and brought his ideas with him to the Australian Museum. By the early 1860s he was working with taxidermist Henry Barnes to take the museum's first photographs.

In the hands of a canny colonial scientist like Krefft, photography became another tool he could use to enhance his own scientific authority and to reframe traditional natural history for his own (as well as his museum's) purposes. Krefft's quest was to make Australian natural history 'scientific', and he used photography as an avenue for circumventing – and sometimes subverting – the social authority and political power of the Australian Museum's overbearing trustees. Unlike most of the trustees who directed his work, Krefft was self-taught and the very model of the practical, field-based, artisan scientist.

With photographs of the museum's specimens, Krefft could trade his knowledge – and provide his own preferred interpretive framework for the objects within the museum, as important scientific specimens and not just as public entertainment. His deployment of photographs of the institution's work and specimens is a key part of the expression of his own scientific identity.

It is the series of fossil digs at Wellington Caves in 1866, 1869 and 1881 (detailed in chapter two) that cemented the usefulness of photography for the museum, and for Krefft. In 1869, money was approved for the purchase of a new camera lens, and £15 set aside for photography and lithography of 'the most interesting of the specimens recently brought by the Curator from Wellington'.[13] The extensive series of photographs taken by Krefft and Barnes on those digs travelled around the world, bringing Australian fossil discoveries to scientific debate in London, New York and Paris.

Back at the museum, throughout the 1860s photography became embedded as a useful tool for documenting both newly arrived specimens and completed taxidermy. Photographs were sometimes offered in return for donations, and in scientific exchanges images of specimens could be substituted for the real thing.

EMBEDDING PHOTOGRAPHY IN MUSEUM PRACTICE

Following on from Krefft's rather individual deployment of photography and his freewheeling, personalised administration of the museum, it was the next curator, Edward Ramsay, who from 1874 to 1894 solidified the presence and processes of museum photography so that it became part of the institution's standard practice. Ramsay, who was the first curator to have been born and educated in Australia, oversaw a period of relative calm and steady growth after the energy and tumult of the Krefft years. Unlike Krefft, he generally had the trust of the gentlemen scientists to whom he reported – and was not as subject to their micro-management. Ramsay headed up a far more autonomous (and less rancorous) regime, typified by a focus on efficient administration, staff stability and collection-building through exchange, donation and purchase. Collections were better managed, for both research and display, with the implementation of disciplinary divisions, item registration and organised documentation of acquisition, exchange and purchase of specimens. Importantly, scientific staff numbers also increased.

Photographs were part of managing the collections within the museum, and they became an official element of the exchange program too in 1876. Although photographic albums had been kept for many years, this had been done in a haphazard way, and they seem to have been arranged for the most part chronologically. Organised photographic albums (discussed further in chapter three), arranged by animal groups, began to be kept in 1879.

Photography could also help frame the Australian Museum for the world. Ramsay's vision for the museum's natural science was cosmopolitan and collective, and at International Exhibitions in Europe, America and Australia across the 1860s, 1870s and 1880s, the museum's photographs were used in public displays alongside taxidermied specimens to represent the institution's modernity, progress and scientific status, as well as its holdings.[14] London's 1862 International Exhibition displayed 600 photographs from all the Australian states, and at the 1880 Melbourne International Exhibition – said to have been visited by almost 1.5 million people – the Australian Museum display included 'three hundred specimens of fish preserved in spirits, with photographs taken from the finest living specimens'.[15] Interestingly, although there are occasional mentions of individual photographic prints being used as part of the Australian Museum's animal exhibits, there is no evidence for their widespread use in public display within the museum.

Opposite: Jacky Winter, *Microeca fascinans*. Museum ornithologist Alfred North discovered these two Brown Flycatchers in a nest in an apple tree near his Chatswood home in Sydney in 1892. He carefully transported the tiny birds to the museum to be photographed, and returned them safely to their nest and their anxious parents the same afternoon.
Photographer: Henry Barnes, jnr

Although they had long been aware of the benefits of photography to themselves and the Australian Museum, by as late as the mid-1880s the trustees had still not adequately funded its operations or expansion. Initiatives for the purchase of new photographic equipment and the establishment of a photographic studio had all been led by museum staff. Like Krefft before him, in 1881 Ramsay was still using his own personal camera equipment (camera, various lenses, stand, case, bellows, glass plates and trays), which he had brought with him to the museum five years earlier. He now requested that the museum purchase the kit as official museum property, which was duly done (although the transaction took another two years to be finalised):

> The whole of the photographs taken during the last five years have been taken with these instruments, they have never been out of this institution or used other than for museum purposes. I find I have no time of my own to carry out photography and beg that the Trustees may buy the outfit which I offer at about £10 less than I payed [sic] for it.[16]

Perhaps it is this transaction that marks the full acceptance of photography as part of standard museum operations. By the time the next curator, Robert Etheridge, jnr, was appointed in 1895 (he had worked at the museum since 1887 as assistant in palaeontology and took over when Ramsay retired due to ill health), photography had its hard-won status, usefulness and longevity confirmed with the construction of new cabinets to house the glass plate collection. As Etheridge explained in his Curator's Report that year:

> A cabinet has been made on the premises for the reception of the collection of Photographic negatives, with a view of bringing them together. They are now serially numbered, thereby saving endless time and trouble in unnecessary searching when a particular negative is required. A Register will also be prepared. The collection now contains over 1500 negatives.[17]

In 1897, just a few years after this, the museum's first purpose-built photographic studio was erected, with darkroom, printing facility and studio space.

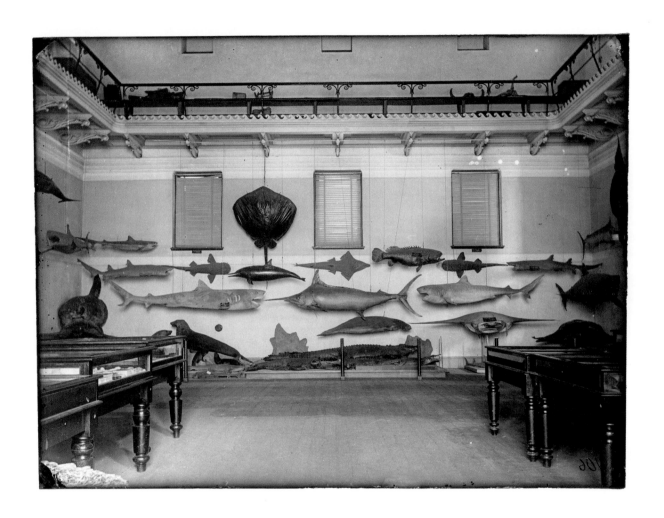

**Above: The Fish Gallery,
early 1870s.
Photographer: Henry Barnes**

TIME, PLACE AND PEOPLE

THE CASE OF KREFFT'S MISSING TYPE SPECIMEN

AUSTRALIAN FRESHWATER CROCODILE, *CROCODYLUS JOHNSTONI*

**Above: Skull of *Tomistoma krefftii*, 1867.
Photographer: Henry Barnes**

**Opposite: Two views of the complete type specimen of *Crocodylus johnstoni*, 1873.
Photographer: Henry Barnes**

Among Gerard Krefft's proudest and most famous achievements was his naming and description of the Australian Freshwater Crocodile, *Crocodylus johnstoni*. Reptiles represented one of Krefft's special interests, at a time when scientific knowledge and description of Australian species was very limited. The series of carefully composed photographs that he and Henry Barnes took of the crocodile at the museum are the only remaining evidence of the complete 'type' specimen. 'Types' are the most valuable specimens in scientific collections, because they are used as the basis for the first published description of a species that gives an animal its scientific name.

Until the late 19th century, the lack of a registration or numbering system for specimens arriving at the museum, along with long distances and large gaps in communication, made naming species new to science involved and confusing.

Krefft had first described the crocodile in 1867, naming it as a new species, *Crocodilus australis*. His only reference was a skull acquired from somewhere near Burketown in the Gulf of Carpentaria, Queensland, which

had been brought to him by a Mr Willam Wood. Wood had informed Krefft that 'the reptiles are plentiful in all the lagoons and waterholes in the neighbourhood of the gulf, and [that] they never grow larger than about seven or eight feet in length'.[18]

A few years later, Krefft sent zoologist Dr John Gray of the British Museum some photographs of the crocodile's skull, noting that it had been damaged during preparation – and apparently forgetting that he had already given it a scientific name.[19] Gray made his own description and proposed naming the species *Tomistoma krefftii*, but the name was never published.

In 1873, Krefft described another crocodile – this time using a complete specimen, including its full skeleton and skin. The animal had been collected by Robert Johnstone (who 'takes a great interest in natural history and if he finds anything rare will forward it to the Museum'),[20] at the upper Herbert River region in Far North Queensland. Johnstone, a sub-inspector of native police, was a keen amateur naturalist. This third description was sent to London and read before the Zoological Society in 1873.

All these crocodiles are in fact the same species, but it is Krefft's third published name and description, endorsed by its official presentation to the Zoological Society, that is the one by which it is still known: *Crocodylus johnstoni*.

Putting the story of these specimens together involved extensive archival sleuthing in 2015 by Australian Museum herpetologists Glenn Shea, Cecilie Beatson and Ross Sadlier in both the Natural History Museum in London and the museum in Sydney.[21] And it is the examination of the photographs that Barnes and Krefft took of the two different specimens that has allowed this fine-grained history of misnaming to be resolved.

The original 1867 type specimen of *Tomistoma krefftii* can no longer be found. Possibly it was photographed by Krefft and then returned to its owner, William Wood. The glass plate negative and a single original photographic print of its skull are all that remain, since the copies of the photographs that Krefft sent to Gray have also been lost.

The 1873 type specimen is still held at the Australian Museum and is in two parts, a mounted skin and a skeleton. Both the mount and the skeleton are missing their heads. It appears that the head was removed from the mount shortly after the species description was finished so that a cast could be made, and that the head was subsequently lost. The series of three glass plate negatives that Krefft printed to send to Gray for the Zoological Society presentation in London, however, show the crocodile with the head still attached. There are also images of the crocodile's side and top views.

The series of five glass plate negatives are now the only – irreplaceable – evidence of the complete type specimen.

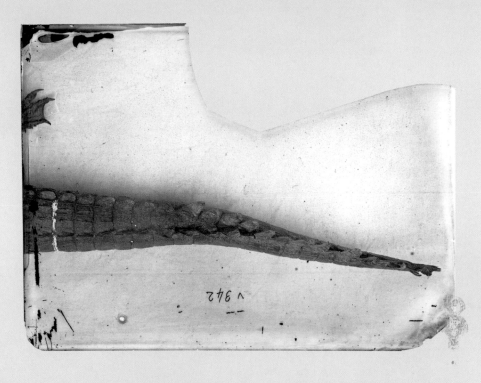

Above: Australian Freshwater
Crocodile, *Crocodylus johnstoni*.
Photographer: Henry Barnes

CHAPTER TWO

THE MEN BEHIND THE IMAGES

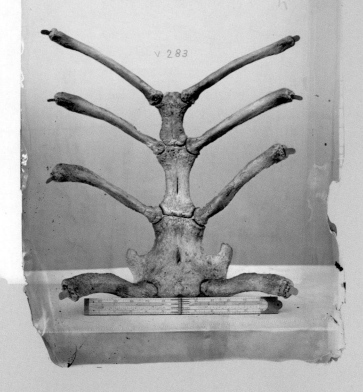

For some time the Trustees have employed a person to take photographs of bones and other specimens of Natural History.

Australian Museum, Trust Minutes, 6 October 1864

Three very different men became the museum's first photographic image-makers. Although none would have regarded himself as a professional 'photographer', each was a trained observer of the natural world, and each had his own practical skills and experience to bring to the new technique and its tools. Of the three, it was Henry Barnes, the museum's taxidermist, who most often set up the specimens for photography, operated the camera, developed the glass plates and made the photographic prints. Curators Gerard Krefft and Edward Ramsay directed operations, stage-managed the production, and made use of the products. Alongside the museum's documentary project, each of these key players sometimes had their own personal goals in mind and in eye as they learnt (mostly by trial and error) to make the most of the cameras, plates, basins, chemicals and papers – as well as the thousands of images they produced over these early decades of museum photography. Although they may not always have got on (it seems that Krefft was particularly difficult to work with), Krefft, Barnes and Ramsay each made an important contribution to the practice, aesthetics and purpose of museum photography.

Unlike a hand-drawn illustration, which is usually signed by its maker, scientific photography can sometimes appear more 'objective' and mechanical than it really is. But like the specimens that make up a natural history museum's collection, these photographs are deliberately and carefully made and used. The Australian Museum is fortunate to have the oldest, largest and most intact documentary archives of any museum in Australia. This unusually rich resource – including administrative records, reports, account books, building plans and of course the photographs themselves – means that it is still possible to not only reconstruct the stories of individual images, but also uncover the stories of the men behind the images.

Page 42: Superb Lyrebird, *Menura novaehollandiae.* Photographer: Henry Barnes

Page 43: Orca, *Orcinus orca*, sternum bones. Photographer: Henry Barnes

Above: Technical staff of the Australian Museum in 1884, including Robert Barnes (standing, second from left), Henry Barnes (standing, third from left), John A Thorpe (sitting, second from left) and Henry Barnes, jnr (sitting on the ground, left). Photographer: Henry Barnes

Opposite: Common Eagle Ray, *Myliobatis aquila.* Photographer: Henry Barnes

GERARD KREFFT: THE MODEL SCIENTIST

Gerard Krefft is the most enigmatic of the characters in the story of natural history photography at the Australian Museum. Born in 1830 to a family of merchants in the duchy of Brunswick, in present-day Germany, he remained something of an outsider in his adopted country of Australia, partly due to his German accent and heritage (although his written English-language skills were excellent). While natural science offered Krefft the chance to move up the social hierarchy – and the position of museum curator was a prestigious one in colonial Sydney – his authority and status was less secure than that of his English or Australian-born rivals. Krefft, largely self-taught, was certainly a talented zoologist, and his scientific record stands as one of the most impressive of any early colonial scientist. He published around 200 scientific articles and described many new species. These included, most notably, the Australian Lungfish (see page 60) and the Australian Freshwater Crocodile (see page 38), as well as many snakes and marsupials.

Above: An unusual studio portrait of Gerard Krefft, from 1864. Krefft holds an Eastern Water Dragon, *Physignathus lesueurii*, in his lap, and an Eastern Blue-tongued Lizard, *Tiliqua scincoides*, sits on the table. Two Diamond Pythons, *Morelia spilota*, can also be seen (one appears to be wrapped around his right arm). Krefft sent the carte de visite to his colleague Albert Günther at the British Museum. Photographer: William Hetzer

Opposite: West Coast Banded Snake, *Simoselaps littoralis*, and Strange's Trigonia, *Neotrigonia strangei*. Odd combinations of animals like this were probably photographed together to demonstrate scale and size. Photographer: Henry Barnes

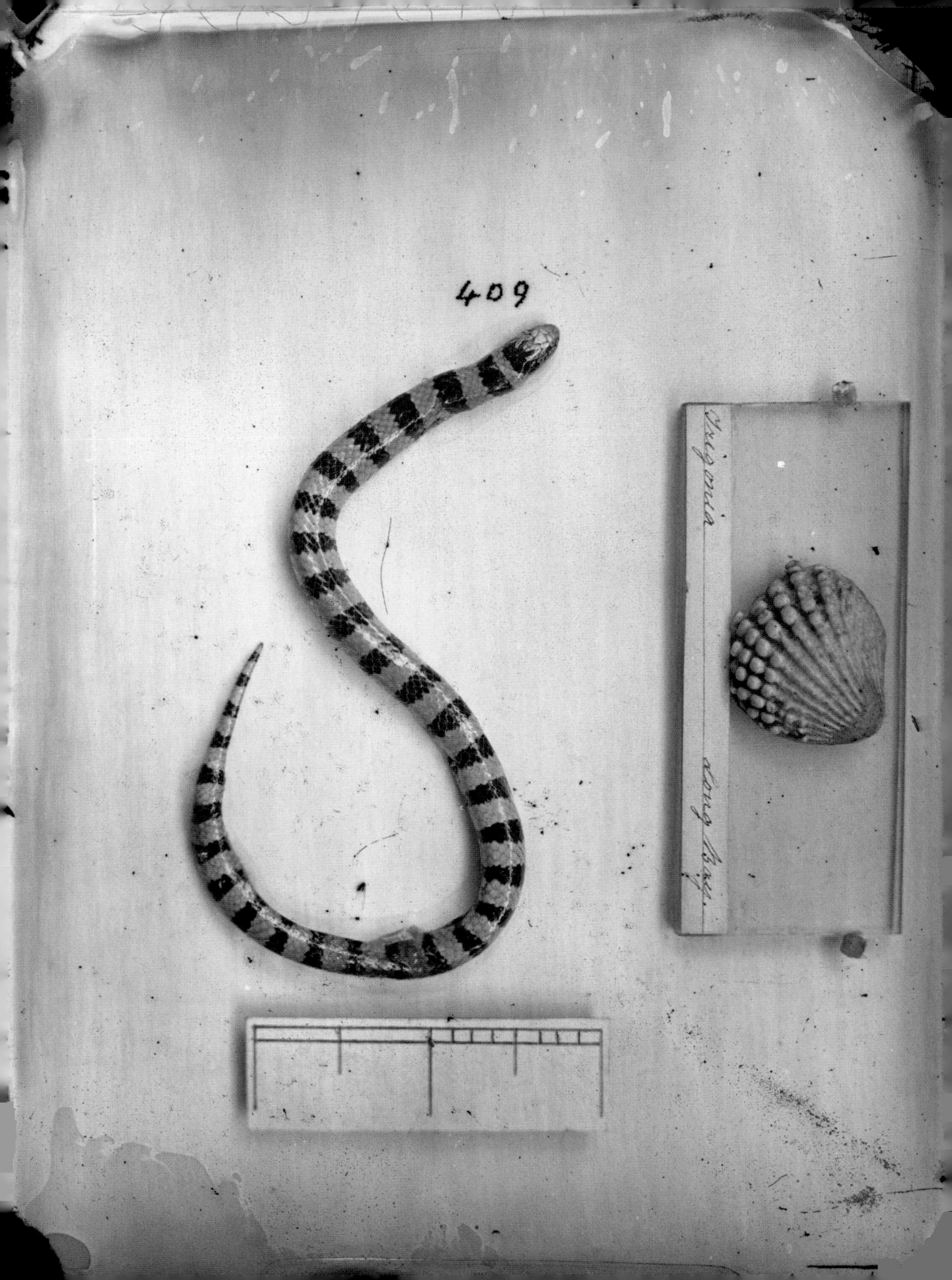
409

Above: Pen and ink drawing of two
Emus, *Dromaius novaehollandiae*,
by Gerard Krefft, unknown date,
possibly late 1860s.

THE MAKING OF A
NATURAL SCIENTIST

The earliest indication we have of Gerard
Krefft's interest in natural history is
characteristically enterprising and
strategic; it shows him with one eye on
the practicalities of earning a living and
the other on a career in science. At the age
of 20, Krefft had emigrated to New York
from his native Germany to avoid the
military draft, and he earned his fare to
travel onwards to join the Australian gold
rush by making and selling hand-drawn
copies of the coloured plates in John
James Audubon's *Birds of America*. After
arriving in Australia he spent five years
on the Victorian goldfields, raising
enough money to return to his private
natural history studies – this time
undertaken at the Melbourne Public
Library, where he made copies of
John Gould's Australian animal
illustrations. Gould's books contained
some of the most luxurious and beautiful
natural history illustrations available in
the country at the time, as well as current
knowledge about Australia's animals.
Krefft's careful copying of the spectacular
large-format illustrations at the library
instilled a good knowledge of Australia's
fauna, but would also have trained both
his hands and eyes in the detail needed
to become a useful field naturalist and
a competent natural history artist.

Krefft had probably met fellow German
naturalist William Blandowski on the
goldfields, and the two men ran into
each other again at the public library.
Krefft was by now 25 years old. The
encounter would shape the rest of his life
and turn him into a working naturalist
when he was invited by Blandowski to
join his government-funded collecting
expedition to the Lower Murray and

Darling Rivers. The exploration party left Melbourne at the end of 1856.

Blandowski had been working as a collector and curator of natural history for a number of years, most recently at the new National Museum of Victoria. He had studied photography and used it on several local collecting trips in 1854, and had even earned a gold medal for his photographic works at the 1856 Victoria Industrial Society exhibition. Blandowski was keen that photography be part of the new expedition, and the group left equipped with a wet plate camera as well as associated supplies of glass plates and chemicals. Although Krefft was highly critical of the equipment, which seems to have been largely useless, this was his chance to learn to use a camera and understand photography's potential. Only one photograph taken on the expedition survives: it was refigured as a drawing captioned 'Portrait of Yarree-Yarree Aborigines based on a photograph by Blandowski, 1857', and appeared in Blandowski's book *Australien*, published in Germany in 1862.[1]

The expedition proceeded to the Murray River, establishing a camp at Modellimin, where the small party based itself for nine months, observing the natural history of the region, collecting specimens and making illustrations. Blandowski was often absent on side trips, or back in Melbourne, and it was Krefft's job to manage the day-to-day camp operations, as well as to catalogue and illustrate specimens. Krefft was meticulous in his work, recording his observations in field notebooks, specimen collection notes and hundreds of sketches of animals and the local Nyeri Nyeri people, who helped him with collecting and with local names for animals and places.[2] Camp conditions

were clearly difficult, as Krefft explained in a letter to Frederick McCoy at the National Museum:

> If you consider, sir, that all these drawings were executed under difficulties, by a hard-worked labourer, under a broiling sun, or in a badly lighted hut or tent with myriads of troublesome insects to contend with, if you consider all this sir, I hope you will give me credit for what I have done.[3]

Both Blandowski and Krefft were well aware of the impact grazing was having on this vulnerable habitat, and Krefft focused on the declining populations of medium-sized mammals, collecting an unprecedented range of species. He also set up what he called 'a small industry' in specimen trade with the Nyeri Nyeri – with animals 'exchanged in the consideration of tea, flour, sugar etc'.[4] Since Krefft's first job in Germany had been as a clerk in a mercantile house, trade, documentation and a commercial approach to natural history collecting would have been his default.

Altogether, the expedition collected 16 000 individual specimens, and Krefft made more than 500 drawings. One album of his drawings is held in Australia, at the State Library of New South Wales. More are archived in institutions in Berlin, Germany, but most have been lost.[5]

The expedition has been called one of the most significant collecting investigations in Australian scientific history – but the zoological results were largely obscured by the controversy generated when the party eventually returned to Melbourne and Blandowski began to present his findings.[6]

Blandowski and Krefft's relationship during the expedition was uneasy. Both

Above: Three views of an Emu,
Dromaius novaehollandiae,
ready for gallery display.
Photographer: Henry Barnes

215

Premolars Molars Right

Dentition of the Thylacine
or Tasmanian Tiger
Lower series.

Left side

IV III II I III II I
Molars   Premolars

Left and right Canine

Incisor I II Right

Left III II I

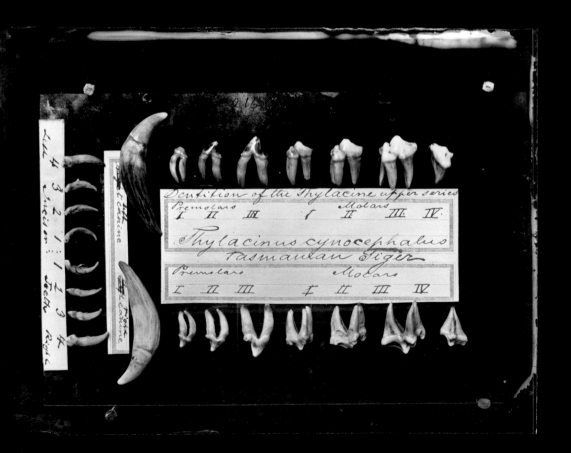

216

Left 4 3 2 1 : 1 2 3 4 Teeth Right
Incisor

Left Canine

Right Canine

Dentition of the Thylacine upper series
Premolars   Molars
I   II   III   I   II   III.   IV.

Thylacinus cynocephalus
Tasmanian Tiger

Premolars   Molars
I   II   III   I   II   III   IV

men were rash and tactless, with an unshakable faith in their own judgment, accompanied by a raw sense of personal injustice – it is not suprising that they did not get along. They finished as enemies. On his return to Germany, Blandowski used Krefft's field sketches of ordinary Aboriginal life as a source for the illustrations in his remarkable publication, *Australien*, which tells an intertwined story of the natural and cultural histories of the Murray. Krefft does not receive a single mention in the work.[7]

Krefft's own entertaining, if self-serving, narrative of the first four months of the expedition shows the difficult, uncompromising side to his character, and perhaps presages his sad downfall from the curatorship of the Australian Museum only fifteen years later. Along with poetic descriptions of the Australian landscape, a sense of real affinity with its animals, and moments of shared understanding with his Nyeri Nyeri companions, the account shows that Krefft was frustrated by his workload and the lack of recognition for his labours, resentful of his lowly status, and unhealthily focused on his rapidly souring personal and professional relationship with William Blandowski. Krefft could not find a publisher for his account in his lifetime, and the manuscript remains unpublished, held at the State Library of New South Wales.

Back in Melbourne, Krefft was employed to create 2000 registrations for the collection at the National Museum. His scientific insights, along with the specimens he collected and the lists, collecting notes and illustrations he made, became his calling card for a career in natural science, opening doors to international scientific networks on his trip to Europe the following year.

Krefft spent 1858 and 1859 in London and Germany, where he visited family after the death of his father. He also used the opportunity to build his credentials, meeting up with colleagues at the British Museum and the Zoological Society (where he read a paper on birds), and in universities and museums in Germany. Here, he traded specimens that he had kept from the Blandowski expedition, and negotiated an agreement to continue to supply more items to the German Museums Commission. Once back in Australia, Krefft presented letters of recommendation from his many scientific contacts to New South Wales Governor Sir William Denison, who secured him a job as assistant curator at the Australian Museum in Sydney in 1860. A year later, after the death of curator Simon Pittard from tuberculosis, Krefft became acting curator, and in 1864 he was officially appointed to the position of curator.

Opposite: Carefully arranged views of the upper and lower teeth of a Thylacine (or Tasmanian Tiger), *Thylacinus cynocephalus*. These images were probably among the 62 photographs Krefft sent to Richard Owen at the British Museum in 1870.
Photographer: Henry Barnes

Above: Pygmy Sperm Whale,
Kogia breviceps.
Photographer: Henry Barnes

THE RIGHT MAN
IN THE RIGHT PLACE

Australia's geology and unique extinct megafauna (large, often giant, mammals and birds) had fascinated European scientists since the early 1800s,[8] and the Wellington Caves fossil deposits (see page 69) offered Krefft an opportunity to make his mark in an international scientific debate, deploying his unique advantages: access to collecting opportunities and specimens, local contextual knowledge and his new photographic apparatus.

Krefft was an early supporter of Charles Darwin's theories, and he often wrote about them in the series of newspaper columns on natural history he produced for the *Sydney Morning Herald* in the early 1870s. He also corresponded with Darwin, and in his earliest surviving communication with him, dating from May 1872, Krefft alludes to two photographic images of specimens relating to his recent Wellington Caves excavations – thus strategically linking himself with the ongoing debate between Darwin and English palaeontologist Richard Owen over matters of evolution. The response indicated that Krefft had successfully cemented himself both as an ally and as part of the scientific network surrounding the leading naturalist of the day. Darwin wrote to Krefft:

> I have read the article with *great* interest. It would be presumptuous on my part, from want of knowledge, to express any decided opinion with respect to yr conclusion. Nevertheless, it seems to me scarcely possible to read all yr statements & reasonings & doubt that you are correct ... It is lamentable that Prof. Owen shd shew so little consideration for the judgment of other naturalists, & shd adhere in so bigotted a manner to whatever he has said.[9]

Gerard Krefft has become something of a patron saint for the history of evolutionary ideas in Australia, and his strong support of Charles Darwin's theories against establishment resistance makes him a useful figure in explaining their impact and effect. The story of his ill treatment and dismissal from the Australian Museum in 1874 has often overshadowed his scientific achievements and distracted from a more nuanced view of his life and work.

In fact, for many years Krefft had the support of the museum's trustees as he worked hard on expanding the institution's collections through fieldwork, donation and exchange, preparing museum catalogues and working on his own scientific papers, newspaper articles and books on snakes and mammals. He was committed to making the museum more accessible and enjoyable for visitors, and he prioritised better displays – with an emphasis on explanatory labels and improved taxidermy and skeleton articulation techniques. For many years, Krefft also worked at the time-consuming task of reorganising the museum's displays in the grand new gallery spaces of the College Street wing, which had opened in 1868.

The many demands of museum administration kept him preoccupied and fretful, and often frustrated. The new wing, for example, was hardly fit for purpose – the windows were so high that a ladder had to be used to open them, and on rainy days the basement flooded and the skylights leaked, leading to damage to specimens and the growth of mould.

The staff was small. As well as Krefft there were the Barnes brothers (Henry and Robert), collector George Masters, messenger Michael O'Grady, attendant John A Thorpe and cleaner Ellen Gillespie. Most of the staff lived at the museum, along with their children and pets, and there was constant conflict over noise and mess. Krefft – and from 1869 his wife Annie and sons Rudi (born 1869) and Hermann (born 1879) – occupied the museum's ground-floor front rooms, and he kept a live menagerie at the museum. At various times there were snakes in the basement, an aviary in the entrance and a tortoise in the museum grounds

(photographs show its removal to the Tarban Creek Lunatic Asylum – later known as the Gladesville Mental Hospital – in 1870). There was even 'an educated pig from New Guinea' that Krefft described as being so tame 'that it follows me around up the stairs and down the stairs. It is a funny little fellow'.[10]

The *Visitors' Guide to Sydney*, published in 1872, provides a description of the museum's public face towards the end of Krefft's tenure: displays were 'numerous and interesting', with rooms dedicated to human skeletons and large quadrupeds; insects; birds and eggs; Australian mammals; whales, dolphins and other marine life; snakes and lizards; fish; Australian Aboriginal and Torres Strait Islander war weapons; minerals, timbers and gold; explorers' documents and relics; and fossil casts.[11] The museum was well respected, well attended and well regarded. This was acknowledged in an 1871 review of Krefft's book *The Mammals of Australia* (somewhat ironic in the light of later events):

> An inward conviction of the fitness of things soon forces itself on the observant mind when 'the right man is in the right place,' and the state of the Australian Museum, since the curatorship of Mr Krefft, is evidence of the fact …
>
> The alterations and improvements effected since his accession to his present office, have, notwithstanding very tardy pecuniary aid, completely metamorphosed the institution, and rendered it one that will, by its many specimens, and the care and accuracy of arrangement, vie with almost any provincial collections as a school of natural history.[12]

Opposite: This species is now known as the Common Wombat, *Vombatus ursinus.* Perhaps this is one of the photos sent by Krefft to the Zoological Society in London in 1872: 'Dear Dr. Sclater, – I have had a series of photographs made of the different wombats: and as it appears there is still some doubt about certain species, I now enclose copies of them'.
Photographer: Henry Barnes

AN UNFAIR FATE

The museum's trustees regarded it as quite proper to gain certain privileges in return for their unpaid services. While the museum certainly benefited from their donations (trustees' names appear constantly in the published lists of donors), there were clear conflicts of interest. Both William Sharp Macleay and his brother George, as well as James Cox, had large private collections. George Masters (Krefft's assistant) collected on a part-time basis for William Sharp Macleay, and other museum staff including Henry and Robert Barnes and Michael O'Grady all occasionally worked for Cox.

Krefft's job as curator was made extremely difficult – at times impossible – when trustees could manipulate the collections or bypass his authority and give instructions directly to staff. Tactless and blunt, Krefft did not disguise his impatience with the trustees, disparaging them as 'bug and beetle collectors' with no scientific credibility.

The coming clash between the new breed of museum worker – exemplified by Gerard Krefft and the progressive scientific ethos and methods that he championed – and establishment natural history – exemplified by a certain set of the museum's trustees – was perhaps inevitable. Krefft would prove a visible and somewhat tragic casualty.

Photography, which he had deployed so strategically and effectively in his work, played a role in Krefft's downfall

Above: Skulls of the Swamp Wallaby, *Wallabia bicolor* (left and centre) and the upper jaw of an unidentified Bettong, or Rat-Kangaroo (right), posed to show relative sizes. The scale is in inches. Photographer: Henry Barnes

too. On 5 March 1874, detectives searched the museum's workshop and impounded a parcel of 'obscene' photographs that Henry Barnes (who had a sideline in commercial photography with his brother Robert) had hidden under his workbench. The Barnes brothers were selling these photographs from the museum premises, and the curator was accused of abetting their sale. The exact content of the images remains unclear, but according to Barnes they were copies of photographs belonging to a friend of Krefft.

Despite the support of George Bennett and another credible scientist-trustee, geologist William Branwhite Clarke (both resigned in protest at the curator's treatment), Krefft was no match for the group of trustees who wished to see him dismissed from his job and who set up a formal enquiry into his behaviour and alleged maladministration of the museum. Although attempts to implicate Krefft in the scheme did not succeed, after the enquiry it was not the Barnes brothers who were dismissed, but Krefft.

The museum's employees testified against Krefft, with accusations ranging from bad temper to occasional drunkenness, conspiracy to fire taxidermist Charles Tost, neglect of a sunfish specimen (discussed in chapter three) and careless supervision leading to the theft of gold specimens from the museum's galleries in December 1873. Of all the staff, Henry Barnes – who had worked so closely and successfully with Krefft over many years, and had even built a rocking horse for Krefft's young son – was one of the loudest complainants. None of the allegations was proven, but the unified voice of the museum's staff is notable: Krefft's brusque treatment of his colleagues had fatally eroded his support. He had long chaffed at the work museum staff were doing for trustees, which apparently involved more than just collecting and taxidermy. At the enquiry, Krefft accused messenger O'Grady of running a catering business from one of the basement rooms, 'killing and preparing poultry for Trustees' private dinners and banquets and then serving as a waiter at the functions'.[13]

Following his dismissal, Krefft refused to leave the museum. Eventually, in a cruel and undignified end to the impasse, trustee Edward Smith Hill brought in 'two prize fighters', who physically evicted poor Krefft, his family and his worldly belongings from the museum.[14]

Krefft spent the next few years in litigation against the museum's trustees for wrongful dismissal, eventually winning his legal battle and receiving compensation. However, he did not regain his position. He died in 1881, aged only 52, survived by his two sons and his wife (and great supporter) Annie.

Perhaps Krefft's obituary in Sydney's *Evening News* best sums up his predicament: 'If he had been as much at home with men as animals, or could have charmed his trustees as cleverly as he did his snakes, his fate would have been a much fairer one'.[15]

A 'MISSING LINK' BETWEEN FISH AND AMPHIBIANS?

AUSTRALIAN LUNGFISH, *NEOCERATODUS FORSTERI*

The Australian Lungfish, as the name suggests, has a rare ability for a fish – it can breathe air both above and below the water surface. Possessing a single lung, the fish can switch to breathing air at the surface if its aquatic habitat becomes too shallow or when water quality is compromised. The large (up to 40 kilograms and 1.5 metres long), carnivorous and long-living fish (it has a lifespan of twenty-five years or more) is found in only a few Australian rivers, including the Mary River in south-eastern Queensland.

The fish has long been known to the Indigenous Gubbi Gubbi people as *Dala*. Dala is an important figure in their creation stories, reflecting the close cultural relationship between the Mary River and Gubbi Gubbi families and their community, as well as their acceptance of fluidity and change in the natural world. The local Gubbi Gubbi were aware of Dala's ability to breathe atmospheric oxygen long before Western scientists (including Krefft) rediscovered this phenomenon in the 19th century. The area's European settlers were familiar with the fish under the name of the 'Burnett Salmon'.

It was Gerard Krefft who first recognised the importance of the lungfish to Western science, and the description and naming of *Neoceratodus forsteri* is arguably his most important zoological work.[16]

In 1870, he was sent two salted and gutted specimens by his friend William Forster – Secretary for Lands in New South Wales and a leading squatter, with property in Queensland's Burnett region. Krefft noted the fish's unusual teeth, its curious internal organs and its combination of gills and a single lung, and immediately saw its relevance to Darwin's theory of evolution: the lungfish could be a transitional animal, the 'missing link' between fish and amphibians. He did not wait for confirmation from his British peers (as was usually the case), nor did he publish his findings in a scientific journal. Instead he sent an urgent letter to the editor of the *Sydney Morning Herald*, which published the news the next day, 18 January 1870.[17]

Krefft named the new fish *Ceratodus forsteri* in honour of his friend. Its official name has since been changed to *Neoceratodus forsteri*.

At the museum, some of the trustees were sceptical. They disapproved of Krefft's Darwinian theories and were annoyed by what they saw as his self-promotion and presumption.[18] But the 'bizarre', 'monster' fish was a media sensation, and the news ran for months in the papers, reaching across the country to become a matter of national pride.[19] Further inflaming the trustees, in letters to the newspapers Krefft proudly recorded the discovery's impact and the praise and congratulations he'd received from his scientific peers.

Above: Australian Lungfish,
Neoceratodus forsteri.
Photographer: Henry Barnes

Above and opposite: Freshly prepared lungfish specimen, 1870. Photographer: Henry Barnes

Krefft and Barnes had also taken photographs of the fish, and it was this series of detailed images that enabled Krefft to again bypass the museum trustees, as well as the usual colonialist zoological communication channels and descriptive practices. Within only a few months, by April 1870, Krefft's scientific paper 'Description of a Gigantic Amphibian', discussing the 'wonderful new beast from Queensland, which is certainly one of the finest Zoological discoveries of the period', had been read before the Zoological Society in London.[20] The prints Krefft and Barnes made of their fish photographs were quickly distributed to Tasmania, where they were shown at the Royal Society, and to London, where Krefft sent copies to his friend Albert Günther – another expatriate German, who was the fish expert at the British Museum. For 'upstart' Krefft it was these photographs that secured both the fish's (and his own) evolutionary notoriety and the primacy of his description and naming.

Specimen hunting began in earnest after Krefft's discovery was published in the *Sydney Morning Herald*, and in August 1870 the Australian Museum despatched the staff natural history collector, George Masters, to Queensland – together with half a ton of 'spirits of wine' for on-the-spot preservation – with the goal of collecting 200 fish. He returned with a more modest 19 new specimens for study, along with photographs of the Aboriginal people who had helped him find and catch the fish, posing with the specimens.[21]

HENRY BARNES: ARTISAN AND TECHNICIAN

Henry Barnes was born in Sydney in 1838 and joined the Australian Museum in 1859, when he was 21 years old. He spent his entire career there, retiring in 1897 and dying just a year later.

Henry Barnes' handprints are everywhere at the museum – not just in the photographic collection but also in the arrangement of early displays, and in the hundreds of animals he worked on as an articulator, taxidermist and preparator. He learnt all these essential museum crafts (discussed further in chapter five) on the job; these skills were mostly rare or unknown in late-19th-century Sydney, and there were few handbooks available.

As with many of the artists, craftspeople and artisan naturalists (including a few rare women) who worked behind the scenes at the museum, Barnes' efforts went largely officially unnoticed unless as numbers in reports. His work represented then, as now, a seamless part of the everyday labour of displays, specimen collection and production, exhibitions and research that make up modern museum practice.

Barnes' ability to find practical solutions to museum problems was boundless. He began his career by tackling the handiwork of articulation, which requires manual dexterity as well as creativity and improvisation. He even invented his own special tool (a drill) for the task.

Within a few years, he had become the museum's main taxidermist and primary cast-maker, as well as a photographic assistant. Krefft declared him 'a perfect genius of taxidermy' for his skills in whale articulation.[22] By 1864, his cast-making skills were good enough that he could make plaster casts of a giant *Diprotodon* skeleton. These were used for a display at the museum, and for an exchange with the National Museum, which owned a rare Pig-footed Bandicoot specimen (*Chaeropus ecaudatus*) that Krefft wanted.

In this same year, Barnes and Krefft had also become confident enough in their photographic productions to begin exchanging photographic prints with Richard Owen at the British Museum. In 1866 and 1869, Barnes accompanied Krefft to Wellington Caves, where they experimented with indoor and outdoor photography. In 1870, Barnes received a bronze medal for an 'Educational series of Natural History specimens' at the Intercolonial Exhibition in Sydney, and he almost certainly took the photographs for Krefft's Highly Commended entry of 'Photographs of specimens in Australian Museum' at the same exhibition.

It is notable, however, that when Henry Barnes and the museum's other three 'artisan' employees (two taxidermists and a carpenter) were bold enough to petition the trustees for a pay raise in 1868, they were given no encouragement to consider themselves unique or valuable. These experienced technicians wanted their salaries to reflect their skill level and relative status within the museum: 'Being Artisans we naturally look for a higher rate of salary than the messenger whose position and duties require no skilled labour'. Their polite letter to the trustees

Opposite: Henry Barnes became the museum's most expert and trusted articulator and cast-maker. He often had to repair and infill the museum's fossils – such as these *Diprotodon* bones, which he reconstructed and set up in the 1870s.
Photographer: Henry Barnes

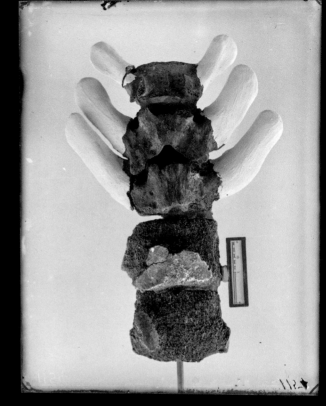

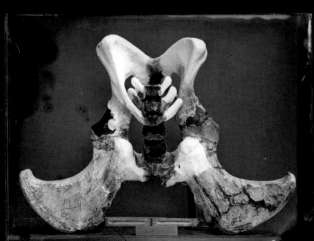

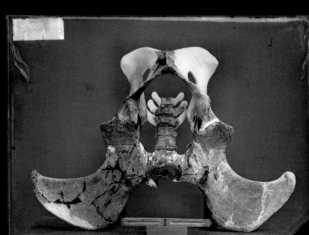

is firmly marked 'moved and seconded, not to be entertained'.[23]

Barnes worked for four curators – Simon Pittard (1860–1861), Gerard Krefft (1861–1874), Edward Ramsay (1874–1894) and Robert Etheridge, jnr (1895–1919) – and was relied on by each of them. Working at the museum became something of a tradition for the Barnes family. Henry, his older brother Robert (who joined in 1868) and son Henry Barnes, jnr (who joined in 1878, aged 16) together racked up more than 120 years of museum service; Henry Barnes, jnr, worked at the museum until 1913.

Ramsay and Barnes seem to have had a particularly close and trusting working relationship, and Barnes held an unusual level of responsibility for his area of the museum's operations. By 1874, because of increased demand for Barnes' services, John A Thorpe had been taken on as senior taxidermist, allowing Barnes to specialise solely in articulation, cast-making, model-making, and displays. It was Barnes who arranged the Australian Museum's displays at the Sydney International Exhibition at the Garden Palace in 1879. In 1880, Ramsay sent Barnes to Melbourne to make casts of fossils in the National Museum. In 1881, Ramsay and Barnes set off to collect fossils at Wellington Caves, with the curator soon returning to Sydney – leaving Barnes in charge of the complex excavation for the next five months.

Barnes also turned his museum skills to personal gain. He is listed in Sydney's *Sands Directory* of businesses as a commercial taxidermist through the

Left: Southern Saratoga, *Scleropages leichardti*. The dried skin of this fish, along with this photograph of the fresh specimen, was sent by Krefft to Albert Günther at the British Museum in 1864. Günther named the fish after Dr Ludwig Leichhardt, who had brought the fish to the Australian Museum in Sydney from the Burdekin River, Queensland. Photographer: Henry Barnes

Departure of N.S.W. Contingent to the Soudan
— 3. March 1885.—

Above: Henry Barnes took museum photographs and commercial images, and also photographed for his own amusement. This photograph from his only surviving personal album shows crowds watching a military parade outside the museum in 1885.

1870s and, like other museum staff, he did extra work for the trustees, in particular James Cox. Barnes sometimes produced commercial photographs too. In 1873, for example, he captured the funeral procession of the statesman William Charles Wentworth from the museum roof, and sold the photographic prints.

After Krefft's removal, Henry Barnes would remain a trusted assistant at the museum for a further two decades. In 1897, the year he retired due to ill health, the museum finally opened a separate 'spirit house' (for wet zoological specimens stored in jars filled with alcohol), a new taxidermy workshop and a photography studio. His departure was duly noted in that year's Annual Report: 'Mr Barnes was an excellent workman, and most of the specimens of skeletons at present on exhibit stand as a lasting monument to his skill. His place has been filled by the promotion of his son, who was trained under him'.[24]

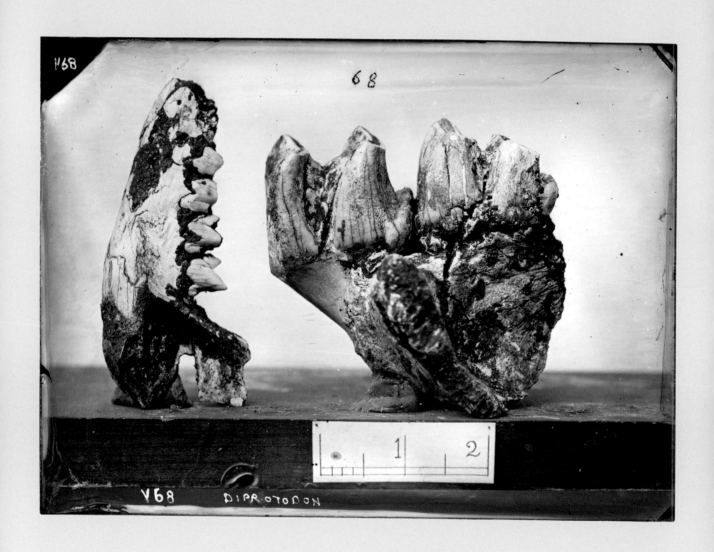

Above: *Diprotodon* teeth
excavated from Wellington Caves.
Photographer: Henry Barnes

AUSTRALIA'S MEGAFAUNA REVEALED TO THE WORLD

WELLINGTON CAVES, NEW SOUTH WALES

The idea of giant wombats 3 metres long, 2 metres high and weighing 2500 kilograms roaming around the outskirts of what would later be the town of Wellington, in central-western New South Wales, was an awe-inspiring thought 150 years ago – just as it is today. When Gerard Krefft began a series of excavations at Wellington Caves (about 8 kilometres south of Wellington) in 1866, revealing fossilised remains of these marsupial giants among large quantities of sequentially deposited extinct Pleistocene megafauna, he knew the site could be an evidential goldmine.

The scientific world was still grappling with Charles Darwin's revolutionary theory of evolution, and the thorny issue of god versus nature had by now captured the public imagination. Krefft's scientific studies had already inclined him towards Darwin's view, but he had been battling fierce resistance from some fellow colonials: most of Sydney's gentlemen scientists saw the evolution issue primarily in theological terms, and they had not yet been swayed.

Emerging in the context of the evolution debate, the Wellington deposits – characterised by fossilised remains of long-extinct marsupial giants found with their more modern relatives – appeared remarkable. They seemed to demonstrate a sequence of faunal evolution, and appeared to illustrate Darwin's law of succession of types. Scientists at home and abroad, Krefft judged, must be tantalised by this kind of empirical evidence. But to be persuasive, its presentation would need to be carefully managed. Coming from the outposts of empire, in the custody of a fledgling colonial institution, it would need to be seen to be believed – and then, Krefft hoped, it would be 'the means of bringing to our shores many scientific men ...'[25]

THE BRECCIA DEPOSITS

Wellington Caves – referred to by local Aboriginal people as Mulwang – had been known to Europeans since the 1820s. Occurring in low hills at an elevation of about 50 metres, the formation of five main caves was first officially explored by Surveyor-General Major Thomas Mitchell in 1830, at which time a collection of fossils was retrieved and swiftly despatched to Richard Owen at the Hunterian Museum in London. Mitchell, a Fellow of the Geological Society of London, was anxious to have the bones examined by the leading scientists of the day, and Sydney at that point had no-one with such a reputation.

News of the deposits was disseminated in British and European circles, and Charles Darwin and prominent Scottish geologist (and later pro-evolutionist) Charles Lyell took note. There was a definite similarity between the ancient 'ancestors' and what appeared to be their still-living relatives – interesting, but more investigation was obviously needed.

In 1866, Gerard Krefft set off from the museum on a 360 kilometre trip to the limestone caves of Wellington with 'a conveyance, two horses, driver, and

Top: Panorama showing the
entrance to Wellington Caves.
Krefft is just visible in the
centre, holding a scale.
Photographer: Henry Barnes

Above: The campsite
at Wellington Caves.
Photographer: Henry Barnes

two men ...'[26] The intrepid curator and his two museum assistants, Henry Barnes and Charles Tost, also had on board what he suspected would be a vital part of this scientific investigation: a heavy load of photographic equipment.

When they arrived, Krefft and Henry Barnes attempted to photograph the magnificent interior of the caves using the bright artificial light generated by burning magnesium powder. Conditions weren't favourable though; and instead, after battling 'a myriad of flies which penetrated everywhere',[27] they finally managed to capture five views of the valley outside and the bold limestone rocks in the neighbourhood of the entrance.

Moving from the main cave into two smaller caverns beyond, Krefft and his men spent the next few days digging laboriously in the dry, granular breccia: 'the tedious task of examining great heaps of red dust is difficult to conceive, as the fine particles of the deposit, rising in clouds at every movement of the body, often extinguish the candles, and make breathing difficult'.[28] Eventually, though, an impressive haul of some 1400 pieces of fossilised bone – representing reptiles, birds, bats, rodents and mammals – was packed and forwarded to Sydney.

A selection of these newly unearthed fossils received their first international viewing at the Paris Universal Exhibition in 1867, prompting an intrigued Richard Owen to request that the Colonial Secretary of New South Wales grant £200 for further exploration of Wellington Caves. This was forthcoming, and in 1869 Krefft – accompanied by Henry Barnes and Dr Alexander Thomson, reader in geology at Sydney University (and amateur photographer) – launched another expedition.

This second trip produced an even greater trove of specimens than the first. Fifteen days were spent hauling, sifting and washing '250 tons' of loose, red breccia. Krefft reported that 'We obtained many valuable and rare specimens, some quite new to science ...'[29] The Australian Museum Report of 1869 detailed a list of 2100 Wellington specimens.

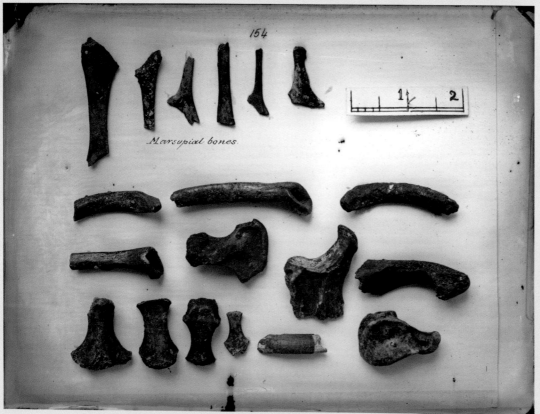

Left top and bottom:
Marsupial bones from
Wellington Caves.
Photographer: Henry Barnes

Opposite top: Fossil
remains of rodents from
Wellington Caves.
Photographer: Henry Barnes

Opposite bottom: Fossil
mammal bones from
Wellington Caves.
Photographer: Henry Barnes

TIME, PLACE AND PEOPLE

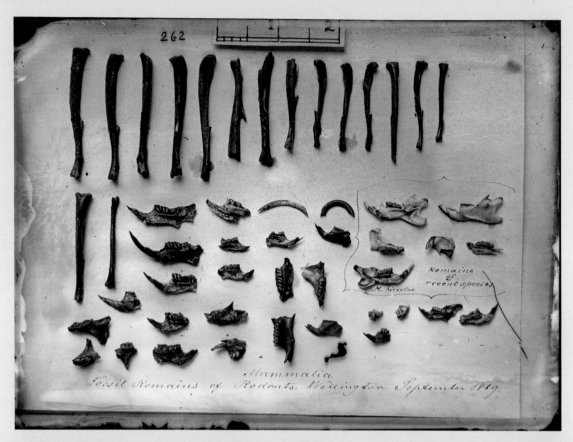

Mammalia
Fossil Remains of Rodents. Wellington September 1869.

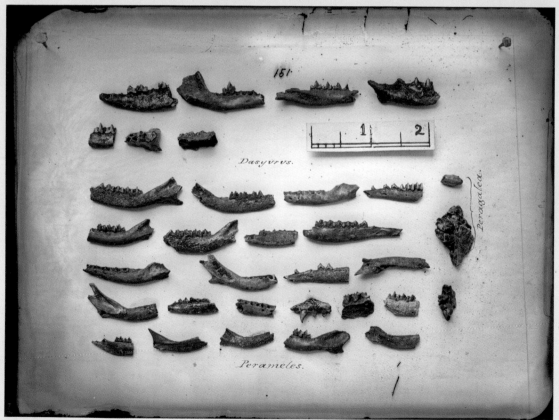

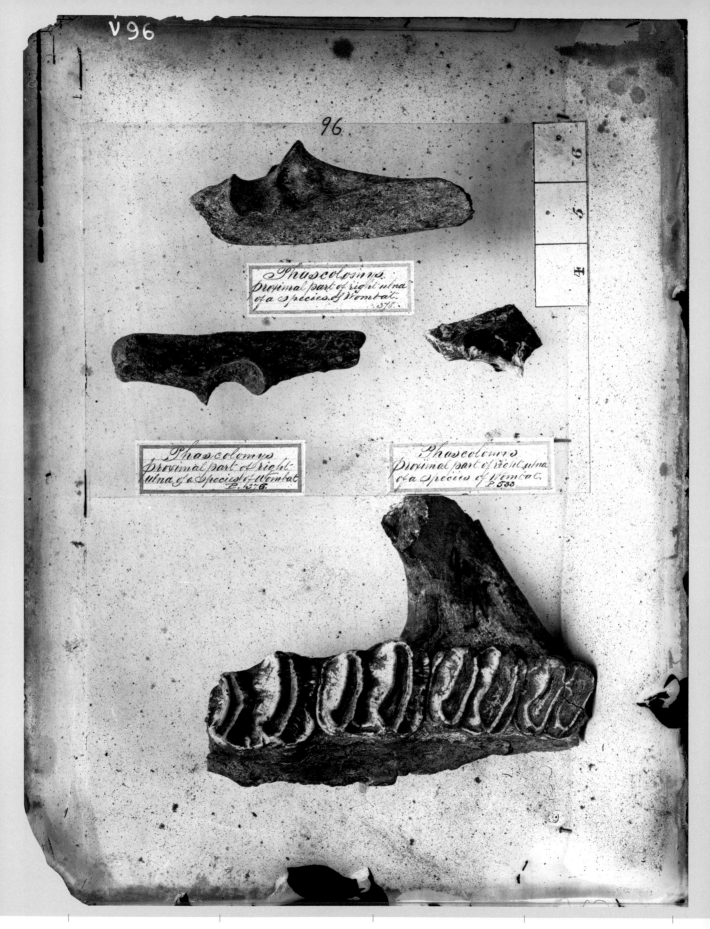

PHOTOGRAPHING THE EVIDENCE

Significantly, many of the bones collected by Gerard Krefft in 1866 and 1869 were never despatched to England. They were instead kept at the Australian Museum, where they were examined by Krefft and others, and photographed.

Krefft was conscious of the need for a colonial naturalist to be especially careful in recording his working procedures and observations in order to make any real impression on imperial science. And he now gambled on the power of the photographic image to help push his case.

Views of the caves and surrounds were photographed on both trips while the field work was underway. Much of the photography of the specimens themselves, however, was undertaken not in situ but back at the museum, the items being carefully arranged against a blank background with a visual scale of their measurement included. Richard Owen was sent a detailed report of the investigations in May 1870, accompanied by duplicate specimens, casts and 62 photographic plates. Lacking many of the physical specimens for study, Owen was sometimes forced to draw conclusions from the photograph alone, and on at least one occasion politely declared himself 'puzzled' by what the image actually showed. Having been sent a photo of a piece of jawbone of an extinct *Nototherium* (giant wombat) by Krefft in 1870, Owen struggled to make out precisely which part of the jaw it represented, commenting, 'You see the difficulty of dealing with Photographs only'.[30]

Through the supply of selected images to key correspondents, Krefft aimed to provide mostly accurate, indisputable evidence of his fossil finds, inviting international interpretation and comment – without having to relinquish the collection itself. He remained in charge of the 'physical' body of evidence, and to some extent attempted to mould the flow of intellectual discourse by means of the strategic provision of relevant images.

A SIGNIFICANT SITE

Following his departure from the museum in 1874, Krefft's important legacy of providing detailed photographic evidence of the caves finds continued. The Australian Museum carried out further excavations at Wellington Caves in 1881, and the results of this work, along with Krefft's earlier findings, were included in an 1882 Parliamentary Report entitled 'Exploration of the Caves and Rivers of New South Wales'. Accompanying this report were detailed illustrations, maps and 14 pages of photographic plates.

Excavations at the caves have continued periodically into recent times, further illuminating the significance of the Wellington site – the first known source of marsupial fossils in Australia and the type locality for several fossil mammal species. Of the 58 marsupial species listed from the site, approximately 30 are now extinct throughout Australia, and another 12 no longer inhabit the Wellington region. The photographic plates produced with such industry and foresight by Krefft and Barnes represent some of the earliest images of Australia's unique megafauna remains.

Opposite: Fossil wombat bones from Wellington Caves. Photographer: Henry Barnes

EDWARD RAMSAY: A CAREFUL MANAGER OF PUBLIC SCIENCE

Edward Ramsay transformed the Australian Museum. With support from the trustees and increased funding from the New South Wales government, he greatly expanded the museum's collections and was the first curator to have a staff of specialist scientific assistants. He was a keen amateur ornithologist in his younger years, and his career in natural science was promoted by his connections among Sydney's close-knit establishment networks, especially with William Sharp Macleay. In 1874, at the age of 32, he was appointed to immediately succeed Gerard Krefft as curator.

The appointment was in fact illegal, as Krefft had been unfairly dismissed; and therefore Ramsay's appointment was not formally confirmed until 1876. Krefft and Ramsay had known each other since 1860, when young Ramsay had begun to donate specimens to the museum – Krefft disliked him, describing him as 'an enemy of mine of long standing on account of my refusing to purchase the rubbish he used to offer'.[31]

If Ramsay's career was boosted by his connections, it is also marked by his affability and sociability, and particularly by his talent for administration, delegation and operational efficiency. For the museum, it was an era marked by prosperity, an outward-facing confidence and the growing stature of museum science. As curator, Ramsay was not an active field collector (with the exception of his involvement in continuing excavations at Wellington Caves),

but he was an engaged, useful and valued member of Sydney's civic and scientific clubs and societies. He attended meetings, acted in official capacities and facilitated the growth and development of the Entomological Society, the Philosophical Society, the Zoological Station at Watsons Bay, the Royal Society and the Linnean Society. He was a government delegate to a series of International Exhibitions.

It was also a period of consolidation and delegation for museum photography. Although Ramsay was himself a keen photographer (he had his own camera equipment when he joined the museum in 1874, and his son JSP [Jack] Ramsay became a noted bird photographer), he was not as actively involved as Krefft had been, leaving the technical details and photographic process largely to Henry Barnes. He was, however, much more interested than Krefft in leveraging photography as a useful, museum-wide (rather than personal) tool, by initiating the indexing and consolidation of the photographic collection in albums.

His local networking paid real dividends for the museum in terms of funding and staffing, and by the time he retired in 1894 the institution had a scientific staff of eight. Ramsay wrote more than 120 scientific papers on ornithology, fish, reptiles and mammals (his 1865 observations of Australian cuckoos are even mentioned in Charles Darwin's *On the Origin of Species*), and he established the *Records of the Australian Museum* in 1890 to communicate and promote the museum's scientific work.[32]

Ramsay also cultivated international networks of scientists and museum colleagues and, following Krefft's lead, he used strategic exchanges to build the collection. In 1883 he spent a year in Europe, attending the Great International Fisheries Exhibition, as well as inspecting museums, aquariums and zoos, meeting his professional peers and establishing exchange relationships. With the availability of increased and more stable funding, he was also able to purchase many more specimens to fill gaps and broaden the collection. In ornithology alone, Ramsay added more than 18 000 specimens.

Ramsay was curator at the time of the great Garden Palace fire of 1882 (discussed in chapter three), when the museum's ethnographic collection – which was being stored in the building – was almost entirely lost. As well as actively collecting after the devastating blaze, he oversaw another expansion of the museum's gallery spaces, with the addition of a third storey to the original Long Gallery.

Undoubtedly, however, Ramsay's most important contribution was the recruitment of scientific staff. No longer could one man, as Krefft had, cover every animal group. The sheer number of specimens now held in the museum made specialisation and systematic cataloguing a necessity. By 1885, there were assistants (later renamed curators) of insects (Sidney Olliff) and zoology (James Douglas Ogilby), soon followed by ornithology (Alfred North) and palaeontology (Robert Etheridge, jnr).

Like Krefft before him, Ramsay lived at the museum for many years with his growing family – this time in relative harmony with other staff and their families who also resided there – and five of his six children were born on the premises. Ramsay was popular with his colleagues, and before he left for Europe in 1883 they presented him with a certificate of appreciation and a pair of binoculars for use in bird-spotting on the voyage.

After his retirement as curator in 1894 he was appointed consulting ornithologist, a position he happily held for another fifteen years. Robert Etheridge, jnr – who succeeded Ramsay as curator – wrote in his 1917 obituary that 'He was known amongst his intimates as a man of most genial manners, kindness of heart, and possessing a rich vein of humour'.[33]

Right: Three views of the nest and egg of the Green Catbird, *Ailuroedus crassirostris*, collected by ornithologist Alfred North near the Tweed River, northern New South Wales, in 1891.
Photographer: Henry Barnes

CHAPTER THREE

MAKING AND MANAGING THE COLLECTIONS

All packages should be addressed to the
AUSTRALIAN MUSEUM, Sydney, where
any reasonable charges for carriage or
freight will be paid.

Gerard Krefft, *Catalogue of Mammalia in the
Collection of the Australian Museum*, 1864

In the 1850s and 1860s, public enthusiasm for natural history was reflected in the steady opening of new galleries and continuing crowds at the Australian Museum – and in renewed commitment to collecting. From the late 1860s, instead of offering just the 'rare and curious', the museum's ambition and mission was to become encyclopedic – just like the British Museum, which was the model and guide. It would be a place where 'rational amusement, combined with instruction, is offered to the mass of people, and where students have every opportunity to examine and study the specimens of which the Museum consists'.[1] The museum was to give the people of New South Wales a chance to see representative animals from around the world and compare them to those they might find living in Australia. Collections expanded rapidly as the public responded with donations, the government increased funding for field collecting, and the first specimen exchanges were made.

Page 80: Most likely a snake in the *Vermicella* genus eating a snake from the Typhlopidae family. Photographer: Henry Barnes

Page 81: Egg of the Kagu, *Rhynochetos jubatus*, normally found in New Caledonia but hatched at Mosman in Sydney in 1902. The egg was photographed resting on a sheet of glass in order to eliminate shadows. Photographer: Henry Barnes, jnr

Opposite: Common Wombat, *Vombatus ursinus*, skeleton presented to the museum in 1899. Photographer: Henry Barnes, jnr

THE MIXED BLESSING OF DONATION

Donations offered the weird and wonderful alongside the useful, but were a continuing and valued source of collection and exchange material. In July 1884, for example, a large range of specimens was donated, both living and dead: 'a mongoose and a pig-tailed monkey' (from the zoo); birds, bats and the skull of a Bengal tiger; birds' eggs and alligator eggs; a variety of fish; and an octopus. Donations often came in one by one, and much of the curator's time must have been taken up in processing the animals and doing the necessary paperwork as a steady stream of specimens kept arriving at the museum.

The same donor names appear in the lists year after year, including patrons, trustees and staff – and a perhaps surprising number of women. Helpfully, donations are all listed in the museum's Annual Reports, and until the 1890s donors were thanked publicly in lists published once a month in the *Sydney Morning Herald*. Natural history artists Harriet and Helena Scott, for example, were regular donors, not just of butterflies and moths (which were their special interest) but of a range of other animals that they collected on their natural history excursions. And Gerard Krefft's work on Australian snakes in the early

1860s brought hundreds of new reptile specimens to the museum, many collected by Krefft himself or – he claimed – purchased with his own money. He was also greatly assisted by his donor network, so much so that one room in the museum's basement was set up for the reception of 'snakes, lizards and other reptiles', alive and dead. At one time, his holdings there included a live viper, a death adder and a boa constrictor.[2]

FIELD COLLECTING

Of the changes Gerard Krefft made at the museum, one of his proudest achievements was to extend and enhance the art of in-house collecting. Krefft had learnt his own natural history craft through field work. He knew first hand that in order to study animals, scientists needed not just a record of the environment in which they were found, but an understanding of their habits and behaviour in that environment – gained through field observation. With the help of some modest government funding increases (although finances were still always precarious, and requests for additional funds were made every year), Krefft could send out his own collector and ensure the highest quality specimens as well as the best contextual data and descriptions.

From 1864, energetic staff museum collector George Masters crisscrossed the continent collecting thousands of specimens, including birds, insects, mammals and snakes. In the course of ten years he collected throughout New South Wales and travelled to Queensland, South Australia, Lord Howe Island, Western Australia and Tasmania. Masters also worked for the Macleay family, sometimes concurrently with his museum work; this was a source of considerable conflict with Krefft. Masters left the museum in 1874 to become William John Macleay's private curator, and served as curator at the Macleay Museum in Sydney until 1912. His entry in the *Australian Biographical Dictionary* states that he was 'a splendid shot, caught venomous snakes with his bare hands and was fearless in the bush'.[3] So successful were his collecting efforts that at one stage he is said to have brought in more than half the natural history specimens in the Australian Museum. Another museum collector, Alexander Morton, was sent to New Guinea in 1877, Port Darwin in 1878, the Solomon Islands in 1881 and Lord Howe Island in 1882, and travelled across Queensland and Victoria for the following decade. It was a golden age for museum collecting – before the economic depression of the 1890s put a stop to collection activities. It was not until after World War I that collecting of this ambition and scale was again attempted by the museum.

Opposite: Northern Swamp Wallaby, *Wallabia bicolor mastersii*. This is probably the specimen George Masters collected in 1870 from the Burnett River in Queensland and that Krefft named after him. Photographer: Henry Barnes

V 351

V 348

Krefft, too, continued to collect, even combining his honeymoon with a field trip to the Liverpool Ranges in the Upper Hunter region of New South Wales to excavate newly discovered *Diprotodon* bones. In 1869, he wrote to Dr John Gray, keeper (curator) of zoology at the British Museum:

I confess I have gone and done it [got married]; the best fun was however that nobody found me out for a good while as I was supposed to be the only person living in the Museum. Having made a clean breast of it to the trustees it happened very opportune that some ancient bones were found up at Murrurundi and it was moved, seconded and carried that I should have a honeymoon at the same time to look after the bones.[4]

He left Mrs Krefft at Singleton while he travelled to Murrurundi, where he excavated and packed his *Diprotodon* bones before returning to his patiently waiting wife.

Of course, collecting work was much complicated by the web of private and personal interests it traversed. It was common for curators and field collectors to be on the lookout for specimens to keep for themselves (to onsell or exchange, or for their own personal collections – or those of sponsors and employers). For most of the museum's trustees a clear conflict of interest existed between their own collections and collecting priorities and their sometimes less than altruistic roles and work at the museum.

There is a 'boy's own adventure' flavour to many accounts of the travails of collection: the work was often difficult, arduous, even perilous, and it was not always entirely successful. Indeed, in an unfortunate precedent, the museum's first collector, William Holmes, accidentally shot and killed himself while on a field trip at Moreton Bay in 1831.

TRADING AND NETWORKING

But if the work of museum and local collectors gets the most attention, in reality the largest proportion of natural history material in the 19th century was acquired by the more prosaic methods of donation, purchase and exchange.

Most purchases were from field collectors, natural history traders and local individuals, bought for both display and exchange. A few purchases of high-quality and hard-to-source display specimens were made from international natural history collectors and dealers in Europe, England and the United States.

Official exchanges began at the Australian Museum under secretary George French Angas in 1856, when the museum's Annual Report announced that 'The Trustees have also entered into a system of correspondence with various foreign Museums, in order to effect exchanges of Australian specimens for those of other countries, a measure which will, doubtless, tend materially to the increase of the collections'.[5]

Opposite: Occasionally, Krefft and Barnes photographed live animals. Snakes and lizards were appropriate subjects as they stayed relatively still, which suited the museum cameras' long exposures. At top is most likely the Eastern Brown Snake, *Pseudonaja textilis*. Shown below is probably a Solomon Islands Skink, *Corucia zebrata* – adults grow up to 35 centimetres long, and they are the largest known species of skink.
Photographer: Henry Barnes

Left: Australian Pelican,
Pelecanus conspicillatus. These
two handsome birds were collected
in Sydney in 1877 and 1878.
Photographer: Henry Barnes, jnr

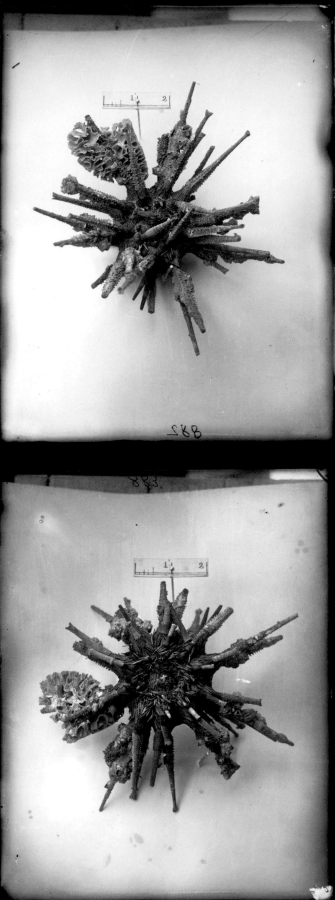

Above: Quartz, siderite and ankerite specimens. It is notable that there are very few mineral, shell or insect photographs in the museum's early scientific photography collections.
Photographer: Henry Barnes, jnr

Opposite: Two views of the sea urchin, *Prionocidaris australis*, found near Sydney in 1885 and described by Edward Ramsay as a new species.
Photographer: Henry Barnes

Exchange was the cheapest way to build the collection – and networks could be built at the same time.

Under what was essentially a bartering system, institutions (as well as a few individuals) would exchange specimens with each other, negotiating to make sure that the transaction was acceptable to those on both sides of the deal. Duplicate bird, fish, shell and mammal specimens from the museum's collections could be traded to fill gaps in the institution's own research and display collections, especially for the collections of European specimens.

In 1885, for example, there were 32 exchange transactions with institutions in Australia (the National Museum in Melbourne, the Melbourne Exhibition Aquarium, the Tasmanian Museum and the Adelaide Museum), New Zealand (the Canterbury Museum), Canada (the Fisheries Exhibition), India, China, Europe (Brussels, Florence, Rome and Vienna), London (the Zoological Society), Cambridge (the Museum of Archaeology) as well as named individuals in France, Scotland, New South Wales, Victoria and Queensland. Every transaction was different, and individually negotiated. Rarely are they like for like: boomerangs were traded for books, birds for fish, and reptiles for minerals. The combination of personal, social and financial accounting around these exchanges is fascinating and complex.

PHOTOGRAPHING THE ETHNOGRAPHIC COLLECTION

Under Krefft's leadership the museum did not actively collect cultural material, and while there are some images of cultural objects in the museum's early photographic albums, they make up only a small proportion of the albums' images (less than 10 per cent). This reflects collecting priorities and relative collection sizes. 'Ethnographic' collections were not of primary importance for acquisition in the Australian Museum's first sixty years, and the cultural collections were small. It was possibly also a matter of practicality and immediacy: fragile, rapidly decomposing and deteriorating animal specimens were chosen for photography over more robust and stable cultural material.

This changed with Ramsay's appointment in 1874. In the late 1870s, Alexander Morton collected cultural objects alongside natural history in New Guinea and the Solomon Islands as well as in northern Australia. Further cultural objects were purchased (along with natural history) from New Guinea collectors Andrew Goldie and Kendall Broadbent. The ethnographic collection began to grow in priority, size and importance. By 1882, the ethnographic collection numbered over 2000 objects, including 359 examples of irreplaceable Australian early contact material. It was stored in the Garden Palace, a purpose-built exhibition building constructed at the south-western end of the Royal Botanic Garden to house the Sydney International Exhibition of 1879. The collection was waiting for the opening of a new Technological Museum, which was to hold the ethnographic collection.

But disaster struck. The entire ethnographic collection was lost in the fire that destroyed the Garden Palace in 1882. Only the catalogue entries and a few precious photographs of these objects remain. Ramsay turned the catastrophe into a collecting opportunity, and within a few short years the ethnographic collection had grown to 7500 objects, mostly by purchase and donation. By 1888, the museum had its own Ethnography Hall featuring objects from the Pacific and Australia. This surge in the importance of ethnographic collecting is reflected in museum photography. From the mid-1880s, images of cultural objects make up a much larger proportion of the museum's photographic endeavour.

Above: Seven stone-headed clubs from New Guinea. These were purchased as part of a larger lot of Pacific cultural material from Sydney-based trading firm Mason Brothers in 1883. Photographer: Henry Barnes

Opposite: Ancestral figures collected by museum collector Alexander Morton in 1878 from the Duke of York Islands in New Britain, Papua New Guinea. The chalk figures were destroyed in the Garden Palace fire in 1882, and this photograph is the only remaining visual record of their existence. Photographer: Henry Barnes

'Ethnography' is the 19th-century term for the 'scientific' study of other cultures. Cultures were classified and studied to reveal similarities and differences between groups and establish typologies of 'race' – the goal being a sort of 'thesaurus of cultures' to match the encyclopedia of animals in the zoology collection. A photographic archive or visual index to the accumulating cultural collection not only helped with practical collection management but also seamlessly incorporated and appropriated these objects into the museum's colonialist knowledge system.[6]

Museum approaches to material cultural collections and their visual recording, ordering and memorialising have changed radically in the past fifty years. Today the Australian Museum respects the inherent rights of Indigenous people to self-representation, including decision-making and interpretation of their cultural material held in its custodial care. For current collection work, source and creator communities are partners in creation, documentation, storytelling, custody, access and care. Reflecting this change, this part of the museum is no longer called 'Ethnology', 'Anthropology' or even 'Cultural Collections' (all previous designations). It is now called 'Cultural Connection'.

Above: Right flipper of a Pygmy
Sperm Whale, *Kogia breviceps*.
Photographer: Henry Barnes

Opposite: Jawbone of an
unidentified dolphin species,
undated. The carpenter's folded
ruler is included to provide scale
(and interest) to the bones.
Photographer: Henry Barnes

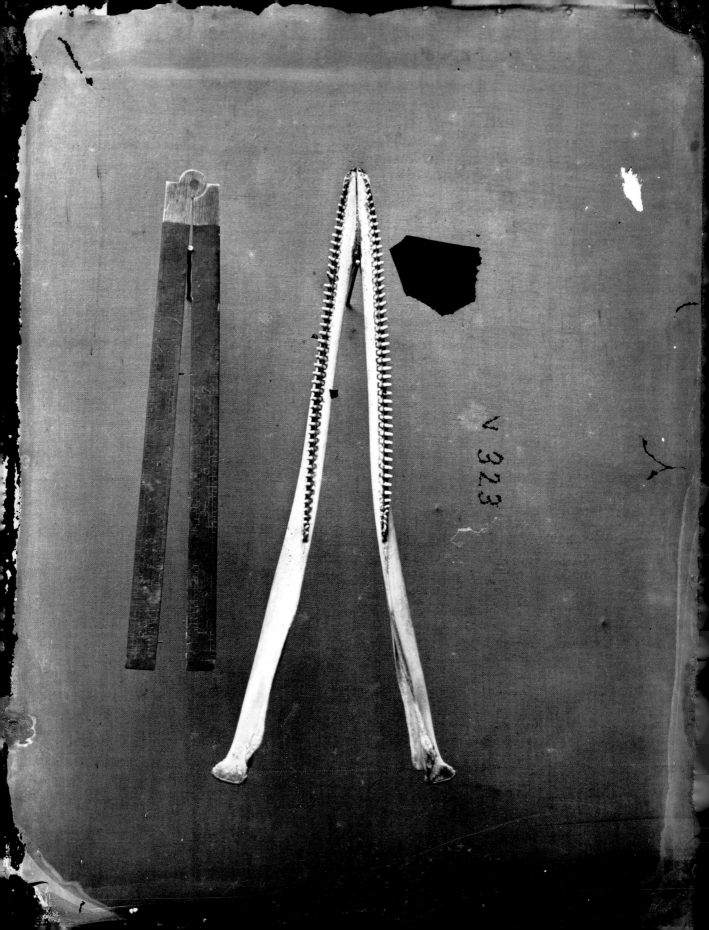

Above: Bump-head sunfish,
***Mola alexandrini*, captured in**
Darling Harbour in December 1882,
and presented to the Australian
Museum by sawmill proprietor
Robert Chadwick.
Photographer: Henry Barnes

SUNFISH IN SYDNEY HARBOUR

BUMP-HEAD SUNFISH,
MOLA ALEXANDRINI

Within just six weeks in late 1882, four rare Bump-head Sunfish were captured in Sydney waters. It was an extraordinary event: these huge, round fish are normally found far out to sea. Occasionally, however, they venture closer to shore.

The largest of the four sunfish was captured inside the harbour by Robert Chadwick and the men working at his sawmill at Darling Harbour, who found it run aground on a mud flat among some moored boats. The monster fish weighed 1116 kilograms, and measured 3.4 metres from the tip of its top (dorsal) fin to the tip of its bottom (ventral) fin.[7] It was as big as a small car. The sunfish was landed on Chadwick's wharf still alive and flapping, an exercise that was said (in an obvious understatement) to have caused 'very considerable difficulty'. However, 'the work was so judiciously performed that the creature was hoisted up [with derrick and tackle] without a scratch being inflicted upon it',[8] as the photograph shown here demonstrates. The man standing proudly next to the sunfish, most likely Robert Chadwick, gives us a sense of its scale – it was reported to be the largest sunfish ever captured in Sydney Harbour.

The Australian Museum was the only place in Sydney that could deal with the preservation of the unwieldy fish, so after 'some fools began to hack it about the head and pectoral fin with an axe and greatly mutilated it',[9] it was lifted by crane onto a truck and sent to the care of the museum. Once the sunfish reached its destination, speed was essential: the fish needed to be prepared quickly before it began to rot and smell and lose its shape in Sydney's summer heat.

The museum's skilled taxidermists were ready for the challenge, as they had previously worked on two other sunfish, captured in 1871 and 1875.

They had not been able to preserve the 1871 specimen, which had already been on paid display in the streets of Sydney for a week before it arrived at the museum. As Krefft remarked at the time: 'there is a wonderful difference between preserving (at a heavy outlay) two tons of bad fish in a lump, and to talk wise over it "in slippers and panama hat" on the Manly pier'.[10] After complaints from neighbours about the noxious smell of decomposing fish emanating from the museum grounds, the sunfish had to be carted off to South Head for disposal, by order of the city's 'Inspector of Nuisance' (whose job it was to keep the city clean, sanitary and safe by inspecting public danger and obstructions, rubbish removal and food safety).

The unfortunate fate of this sunfish became a point of contention in Krefft's dismissal in 1874, with the curator blamed for the delay in accepting it into the museum and for the failure to plan for the fish's immediate preservation. Arguments about the fish spilt into the Sydney press, and they show that by 1871 Krefft already had public enemies. The hostility between Krefft and museum trustee Dr James Cox, who was the go-between in the purchase of the sunfish from a Mr Skinner, is particularly noteworthy.[11]

The 1875 sunfish was prepared more successfully, however, and went on display in the museum's upstairs gallery.

A TRIP TO LONDON

By 1882, the museum's team, this time led by taxidermist John A Thorpe, were experienced and well prepared, with the necessary equipment – and an outside shed – for working on smelly, difficult and oversized specimens like the one found in Darling Harbour. After being examined, measured and described by curator Edward Ramsay, the fish was gutted by making an incision along the dorsal and ventral edges. The skin was preserved, and this specimen became the type for *Orthragoriscus ramsayi* (since renamed *Mola alexandrini*).

The sunfish went to London in 1883 with Ramsay, where it was stuffed and mounted at the British Museum, ready for display at that year's International Fisheries Exhibition alongside hundreds of other Australian fish. Ramsay was the Australian commissioner for the exhibition, and the museum's much-praised exhibits won numerous awards. Rather than bring the surplus sunfish all the way back to Australia, after the exhibition Ramsay donated the specimen to the British Museum.

When that sunfish was brought out of storage in 2016, conservators at the Natural History Museum in London found that it was 'stuffed with the equivalent of 25 large refuse sacks of wheat straw',[12] along with some unusual items such as a broken chair and a scrap of the *Sydney Morning Herald* newspaper from 26 January 1883.

DISPLAY AT THE MUSEUM

The other three sunfish were caught at Manly, Little Manly and Botany Bay,[13] with both of the Manly fish ending up at the Australian Museum (one needed to be towed across the harbour and was nearly lost in rough seas crossing the Heads).

They were made ready for museum display, but the main door of the museum was neither wide enough nor tall enough to provide access for large objects like the finished sunfish. The museum's resourceful preparatory team rigged up a pulley system and the fish was carefully hoisted in through an upstairs window that was hinged to swing open, allowing the specimen to enter the gallery.

Opposite: After being preserved by taxidermists in an outdoor workshop in 1883, the huge sunfish specimen had to enter the museum gallery via the tallest available opening: an upstairs window. Photographer: Henry Barnes

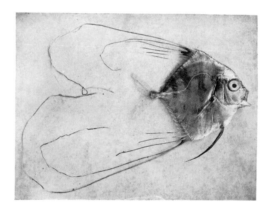

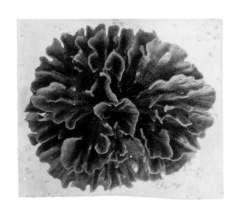

GAINING CONTROL OVER A GROWING COLLECTION

Managing the Australian Museum's animal specimens has always been, and remains, a challenge. They are vulnerable biological material, subject to damage from pests, climatic conditions and light. Until the 1850s, the museum's collections were somewhat chaotic. Everything was kept in cases in the galleries; all the specimens had to be on display since there was no other storage available. In 1859, just after the grand opening of the museum's Long Gallery, it was reported to parliament that:

> the specimens were exhibited in a most incongruous way – everything, instead of being sorted and exhibited in a proper manner, was mixed up in a higgledy-piggledy fashion. Here a fish with bullock's eyes, and there a wild cat with the eyes of a fish. The whole presented the appearance of an old curiosity shop instead of a museum.[14]

KREFFT'S MULTIPURPOSE CATALOGUES

One of Krefft's first, and most enduring, tasks was to establish order in the rapidly growing collections and to make sense of the displays. He complained often and bitterly of the conditions for display within the museum – there was too much light in the galleries, mould growth was left untreated, insect pests roamed freely, and careless visitors sometimes damaged fragile specimens.

In the early part of the 19th century, specimens were not managed as unique objects; instead, they were seen as representatives of a species. The collection history, or provenance, of an individual object mattered less than its correct identification and placement within a systematic hierarchy of animals. The arrangement of cases at the museum reflected this, with animals sorted into biological categories (kingdoms, classes, orders and genera) and sometimes locations (Australian mammals, for example, had their own gallery). Skeletons were housed in a separate Osteological Gallery.

This approach to collections meant that early museum staff saw no real need for individual registration of specimens. As long as the collection remained relatively small, this method could work. Consequently, the first thing Krefft wanted was not a specimen registration system to make internal management of the specimen collection easier, but a published catalogue that could be distributed widely and give his scientific collecting a public face. This was in keeping with his interest not just in placing animals in taxonomic order but in describing their anatomy and physiology as well – and it also reflected Krefft's outward orientation towards his research, his reputation and his far-flung networks. He produced a series of museum catalogues that give a broad

Opposite: Since the original glass plates were occasionally broken or deaccessioned (disposed of) from the collection, the museum's photographic albums contain the most complete record of the entire early photographic collection. Some of these prints are the last remaining images of specimens that no longer exist. However, most of the prints are small, and they are sometimes faded and damaged.

**Above: Juvenile Sumatran
Rhinoceros,** *Dicerorhinus
sumatrensis.* **The skin and skeleton
were presented to the museum by
the Zoological and Acclimatisation
Society of Victoria in 1884. The
skeleton is still in the museum's
collection but this photograph
is the only remaining evidence**

overview of collections but that do not detail the individual collection history of particular specimens and objects.

By the 1870s, the scientific view of specimens had changed, and collections had grown. Krefft's generic approach was no longer philosophically, scientifically or practically possible.

Under Ramsay's more managerial approach, better collection management became the priority. Ramsay recognised that the museum's objects, and knowledge about them, needed to be codified and written down. Species were by now mapped onto a 'type' specimen, and these precious types needed to be carefully and individually documented and managed in secure, safe storage so that they could be used for comparison and research. With this change in perspective and philosophy, the division of the museum's holdings into display and research collections had begun.[15] Specimen collections, research and the knowledge ecology of the museum – including its photographic collections – were splitting off from its public entertainment and displays, and becoming just as important.

Above: The Australian Museum register of photographic negatives. The numbers on the left (which were etched onto the glass plate) are matched up with notes on the image content, usually a museum specimen.

RAMSAY TACKLES THE DETAIL

At a more practical level, the growing collection needed to be individually numbered, labelled and registered to track its acquisition, storage and use. Accordingly, one of Ramsay's first acts as curator was the appointment in 1877 of Edward Palmer, founding secretary of the Linnean Society, to catalogue the collections – a task that would take five years. A series of cataloguers were employed in the decades after Palmer, and the registers were split into separate alphabetical volumes, with a unique letter prefix assigned to each scientific discipline in the mid-1880s.

Unfortunately, the extended and tardy cataloguing process has left some specimens, to this day, without a history. Fittingly, these mystery specimens were allowed to languish in the 'X register', the last volume of the museum's alphabetical system of collection registration.

Fortunately, however, these precious Collection Registers (the entire series runs to dozens of volumes) also provide cross-links between the specimens and any correspondence and official documents about their acquisition (through purchase, exchange or collection). Attempts to track specimens' origins and maintain complete knowledge and control of the collection continue today. With 18 million objects (and growing) to manage at the contemporary museum, it is a never-ending task.

The difference between Krefft and Ramsay can be seen in their distinctive approaches to the museum's growing specimen and object collections – approaches that also define their broader scientific legacies. Krefft collected for his own private research projects, and in his areas of interest. His science had a strong alignment with his self-expression and identity. Ramsay, on the other hand, although he also made a substantial individual contribution to Australian zoology, oversaw a transition to a larger staff and a more measured approach to museum management, and stands for the consolidation of Australian public science through the late 19th century.

Above: Side view of a White Shark, *Carcharodon carcharias*, posed to show the size and shape of its head and mouth. Opposite is the same shark, crudely propped up on a bucket, with rods to hold its mouth open and show off its teeth and characteristic long, pointed snout. Photographer: Henry Barnes

'USEFUL, USELESS, NOXIOUS'

ANIMAL ETHICS

In curator Edward Ramsay's collection of books was a well-thumbed and annotated copy of an 1840 taxidermy manual by William Swainson, a notable British naturalist. Swainson recommends the following list of basic equipment for the field collector:

· A double and single barrel gun, with an assortment of caps, flints, shot, spare screws etc
· Dissecting instruments for opening quadrupeds, birds etc
· Preserving drugs and preparations
· Bottles for containing subjects in spirits, fitted into cases
· Canvas knapsacks
· Corked store bottles for insects, and others for the pocket and for immediate use
· Pins of all sizes
· Boxes fitted with moveable trays for bird skins
· Apparatus for collecting insects
· Chip boxes of different sizes, for small and delicate shells etc
· Knives, scissors, needles, thread etc.

Nineteenth-century natural history field manuals like this one provide confronting details of how to kill animals and what to do with them once they are dead, with only the most cursory regard expressed for the animals' welfare.

Without doubt, 19th-century field collecting formed part of an era of appalling cruelty to animals. European colonists' encounters with Australia's fauna – which they only gradually came to understand and accept – were often hostile and unhappy. In the colonial era, most of Australia's immigrants (including most naturalists) regarded animals in the same way they regarded the other features of the Australian environment: as tools for human use. In their view, animals had neither feelings nor any intrinsic rights, but rather were here for (European) humans to exploit, gather, trade, farm and manage or kill. Both native animals and introduced farm animals were seen as commodities, and their individual welfare carried little economic or ethical weight. Indeed, most animals were treated with complete moral indifference.[16] As the 19th century progressed, colonial folklore about Australia's 'dangerous' endemic animals shaped public perceptions of the hierarchy and value of animals into 'useful' (could be traded, eaten, farmed or used for transport), 'useless' (everything else) and 'noxious' (spiders and snakes).

In the case of field collectors, there seems to have been no limit to the number of specimens killed in the name of trade and science. The Blandowski expedition collected 16 000 animal specimens in just nine months, including hundreds of the small mammal species that Krefft knew were already in decline.[17] Indeed, it is said that Krefft, although a determined and focused collector, did what he needed to do to survive a cold winter on the Murray River and ate one of the last specimens of the Pig-footed Bandicoot, *Chaeropus ecaudatus*, now thought to be extinct. 'I am sorry to confess my appetite more than once over-ruled my love for science', Krefft wrote in his journal. Reports of Australian Museum field-collecting trips from the 19th century regularly referred to birds being shot by the hundreds.

Opposite: Gerard Krefft holding up a freshly caught Leatherback Turtle, *Dermochelys coriacea*, in the museum's grassy grounds. Photographer: Henry Barnes

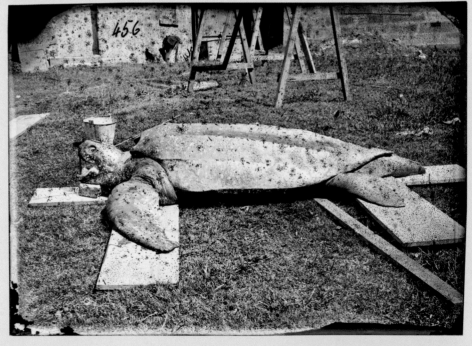

Left, top: The two (headless) men holding the turtle to show off its underside and flippers are probably Robert and Henry Barnes.
Photographer: Henry Barnes

Left, bottom: The huge turtle was carefully photographed from every angle.
Photographer: Henry Barnes

Vivisection, or experimenting on live animals, was also part of the toolkit of late-19th-century natural science.[18] At the Australian Museum, Gerard Krefft experimented by testing different snake venoms on indigenous frogs and lizards, as well as on dogs, cats and goats. He experimented on echidnas too, reporting that they were 'tolerably snake-proof' and that 'it is difficult to drown one, and from eight to ten minutes at least are necessary for the experiment'.[19] However, it was vivisection that brought to public attention arguments about whether scientific knowledge was more or less important than the value of the lives of the animals that were sacrificed (and the pain and suffering they endured). An Act for the Protection of Animals – the first such act in Australia – was passed in Victoria in 1881. The work of the 19th-century anti-vivisection leagues laid the foundations of the animal protection movement of the early 20th century, which in turn led to the development of new concepts of animal ethics and animal rights in the 1960s.

MUSEUM ETHICS

These movements had a profound effect on the practice and mission of natural history museums, including the Australian Museum. All of the museum's current collecting and animal research must go through an animal ethics approval process and comply with the NSW *Animal Research Act 1985*, to ensure that animal projects are 'justified, provide for the welfare of those animals, [and] incorporate the principles of replacement, reduction and refinement'. Permits issued by government agencies are required for all collecting.

SNAKE-BITE EXPERIMENTS AT THE SYDNEY MUSEUM.

Collecting remains a necessary and important part of the Australian Museum's work. Even though there are an estimated 18 million specimens in the zoological collection, new species are still being discovered by museum taxonomists; ichthyologists, for example, discover a fish species previously unknown in Australian waters at the rate of one per week. And more than at any time in the past, collections are providing an essential resource to help answer major scientific questions; relating to biodiversity, for example (which requires study of species from different times, locations and life-history stages), or to environmental impacts including climate change and regional adaptations over time. Photographs cannot provide sufficient detail for this work, although 3D scanning and DNA sampling techniques are changing some collection description practices and may mean that fewer full specimens are collected in future.

Above: 'Snake bite experiments at the Sydney Museum', *Illustrated Sydney News*, **15 May 1880. Animal experiments continued in the cause of science and as public spectacle at the museum into the 1880s. In this case, several canine 'waifs and strays' were captured for the experiment, which was conducted in public under Edward Ramsay's watchful eye. He is seen here kneeling with the dog. A copperhead snake was incited to bite the animal, 'drawing blood and eliciting loud cries from the dog', which later exhibited signs of 'having suffered severely from the poison injected into it'.**

Salmacis Woodsii sp nov

Goniopora tubaria

Patiria crassa

Gomophia tubaria

Phyl australis sp nov

THE PHOTOGRAPHIC ARCHIVE

It is fortunate that almost all of the fragile photographic glass plates that Krefft, Barnes and Ramsay created with such painstaking care more than 150 years ago have survived and are now held in secure, climate-controlled storage within the Australian Museum's archives. A few plates have been broken, and some more were destroyed in the early 20th century in an overzealous cull of 'irrelevant' material, but the glass plate collection is still largely intact. Around 90 per cent of the images in this remarkable archive have survived.

THE 'V NEGATIVE' COLLECTION

The glass plates are the archives' oldest photographic collection, with the first plates dating from the mid-1860s. They are known as the 'V negative' collection. The series contains more than 10 000 plates of varying sizes; most of the earliest plates are 'full-plate', measuring $6\frac{1}{2} \times 8\frac{1}{2}$ inches (165 × 216 millimetres). The last glass plates in the collection were taken as late as the 1940s. Because they are so large, the plates are generally crisp and clear; once digitised and enlarged (either by carefully scanning on a flatbed scanner or with the use of an overhead camera and lightbox), they can reveal not just background information but also wonderful detail of texture and tone.

Unfortunately, before 1893, none of the glass plates were dated, indexed or registered at the time they were taken, and so parts of the history of the making of the collection have been lost. Although albums of prints made from the plates were created from the mid-1870s, it was not until almost twenty years later that the first 1500 glass plates were listed by number, title and photographer. By then, knowledge of the content of some of the images had already been lost and there are some early images for which the museum has no information at all – no title, no date and no photographer. Even after the museum's photography began to be more systematically managed, registration was not necessarily in chronological order and often happened many years after the images were taken. Registration at the time of creation did not begin until 1906, by which stage the collection already held 2500 photographs.

So far it has been impossible to be definitive in dating the earliest museum image. It has taken the dedicated work of archivists, researchers and collection managers to establish confirmed dates for just a selection of images, including many published here for the first time.

Although the early images that are the focus of this book are broadly known to have been created before 1893 – when they were all listed – dating individual examples is only possible through detailed detective work involving study of the photograph's content, which can sometimes be matched to known events, locations, people or specimen acquisition dates.

Opposite: Early photographic album page, showing clear signs of wear and tear, as well as fading where it has been left open and exposed to light.

THE AUSTRALIAN MUSEUM
PHOTOGRAPHIC ALBUMS

Gerard Krefft, Henry Barnes and their successors took photographs for practical purposes, but their photography was never just a technical skill, or just a useful art. Their photographic experiments occurred not only in the context of museum practice, but also within the continuing development of photography as both a technology and an art form.

Photographs always contain a (literal) point of view – far from being merely the product of a 'mechanical eye', they are also an act of individual composition and expression. The images the men took certainly possess aesthetic qualities; some have an ethereal beauty, and many have a quirky charm. So, while certain photographs show the influence of the artistic tradition of hand-drawn scientific illustration in their composition and styling – for example, with smaller fossils laid out in careful grids – others seem to reference the developing tradition of studio portrait photography. And even, perhaps, the new art of the police mugshot, which could be used for identifying, cross-checking and tracking criminals – serving the same basic purpose as the museum's animal portraits.

The museum's photographers did not regard photographs simply as images – they were also *things*, material objects that could be sorted, arranged, kept and even, sometimes, destroyed. Individual glass plates were reproducible as prints, and could then be given, sent and exchanged. The prints were also curated into albums.

The first museum photographic albums were probably made in the mid-1870s, by which time there would have been many hundreds of images to sort and order.

In 1879, at the direction of the trustees, the albums were reorganised, with the aim of making them into more useful working tools: 'On the [motion] of Captain Onslow [one of the trustees], the curator was directed to obtain scrap books in which to mount the photographs taken at the museum, the photos to be [scientifically] arranged'.[20]

There are two types of early photographic album. Presentation albums were made for show and display, with embossed covers, heavy pages and beautiful cursive captions. Volume 19, the 'Fish' album, for example, was sent to the Melbourne International Exhibition in 1880 and was exhibited alongside a selection of the museum's fish specimens. However, the majority of the series of 22 museum albums were 'finding aids' for locating particular negatives. Since most of the museum collections were not registered and were stored in overcrowded cases across the galleries until at least the 1890s, the albums would also have provided a quick ready-reference index and guide to the specimen collection for internal museum purposes. The wear and tear on their bindings and pages is the best evidence for their frequent use.

The first of these practical albums are somewhat disorganised, with proofsheet photographs (prints made at the same size as the original glass plate) pasted in as they were ready, together with other prints that seem to have been added in somewhat random order. The sequence in which the prints were pasted into the photographic albums sometimes provides collection detectives with clues to the chronology of the images, which can, potentially, be matched to specimen acquisition and display dates.

As images multiplied and escaped the capacity of individual memory to recall and locate them (a situation mirrored in general by the museum's rapidly growing collection and knowledge base), the albums had to become better organised and more systematic. And, as photographic equipment improved and became more portable, added to the images that were taken at the museum itself were images taken outside the institution by staff on expeditions and field trips. Photographs came in from external sources, too – acquired through donations, purchases and exchanges among the museum's local and international networks, but especially locally, from donors and amateur photographers in Sydney and New South Wales.

As the museum developed, so did the possible uses for photography and the sophistication of indexes and catalogues. And just as the museum's collections and science split into disciplinary streams, so too did the photographic albums. From the late 1880s, there are separate albums for each natural science collection – and more for ethnology, as well as general museum (non-collection) albums. Some of these were still in use in the 1950s, a historical record of collection growth and change, holding hundreds of pages of collection images. The leather-bound albums are large format and heavy, with up to a dozen photographs on each page. For the browsing non-scientist there are mesmerising repetitions – page after page of images of glossy fish, elaborate nests, birds' eggs, coiled snakes, sea stars and seahorses.

Above: The beauty of the photographic prints and the albums, with their composed pages, delicate sepia tones and otherworldly imagery, is a reminder of how close art and science remain in this unusual photographic archive.

PHOTOGRAPHS AS IMAGE, OBJECT AND INFORMATION

For most of their existence, the albums remained in use within the science and ethnography collections, while the glass plate negatives were held and managed by the photography department – one of the oldest continuing functions of the museum. In the 1990s, the glass plate collection was transferred to the museum archives. There, once the plates had been rehoused to archival standards, the lengthy task of description and, latterly, digitisation began. This move has led to a better understanding of the historical context around the production and use of the museum's entire photographic collection. It has also highlighted photography's importance both as an integral part of museum practice, and as a means of documenting and historicising this practice.

Within an archive within a museum, the nested meanings of photographs and museum photographic processes and uses multiply. In the Australian Museum's archives, the photographic collections are surrounded by the paper collections that can help make historic sense of them: the registers, accession records, accounting books, correspondence, publications and reports that document the way photography was set up, used, formalised and embedded as an important part of the museum's toolkit.[21] And the museum is additionally fortunate – not only does it have the photography collections, it also retains many of the specimens they document. These early images are still useful for recording the history of those specimens, their condition on arrival, any changes in their condition, or treatments they have undergone over time. The images also visualise other parts of the museum's history: the buildings and grounds, display cases and their contents, the arrangement of the galleries, and the difficult conditions under which the staff worked. Used together for historical research, photographs and paper records can each play their part in documenting the co-creation, codification, curation and long-term care of scientific knowledge at the museum.

As Krefft and Barnes understood, photographs exist as both object and information. They don't just reproduce content but are things that are *used*. With a new focus on these unique records of the institution's own photography as a collection in its own right, we are beginning to research its rich content and to understand the fundamental role of the Australian Museum's photographic project in constructing the modern museum.

Opposite: Lumholtz's Tree-kangaroo, *Dendrolagus lumholtzi*. This elaborate group of tree-kangaroos was set up in 1890, and photographed before it was installed in a special custom-made case that was said to have cost the extravagant sum of £13. The specimens were collected in north Queensland, some by museum collector Robert Grant on an expedition in 1889 and others in 1888 by Mr GE Clark from Herberton. Photographer: Henry Barnes

The Tree Kangaroo.
DENDROLAGUS
LUMHOLTZII, Collett.
Loc. Herberton District, Queensland.

HENRY BARNES' DAY OFF

LANDSCAPES AND VIEWS

– *cockatoo* –

Photography was not all work for Henry Barnes and Gerard Krefft. On rare occasions, it could also be a social activity. Krefft and Barnes worked together for fifteen years, and for most of that time they had a comfortable and amicable relationship. Sometimes they spent time together outside the museum, on photographic outings around Sydney to record local landscapes and views. The harbour's breadth and inlets provide the backdrop to some of these views. Or Krefft and Barnes stayed on at the museum on Saturday afternoons (they had Saturday afternoons and Sundays off) photographing local streets, the museum's buildings and grounds, each other, and family members. A collecting trip might be an opportunity to capture the museum's men at work; alternatively, it could provide the chance to capture relaxed, private moments, sometimes with family.

Alongside the thousands of specimen images they took for their work (where the point of view is rigorously objective), just a few dozen of these precious personal photographs remain, pasted into two albums: a presentation album that once belonged to Henry Barnes, and another that was kept at the museum. Mostly, we don't know who the people in these photographs are. It is likely they are Henry Barnes' wife, his brother and his children. But the albums reveal a side to Barnes that his decades of work on the museum's other scientific photographs never express.

For example, on his second trip to Wellington Caves, in 1881 – when Ramsay left him in charge of the dig and its team of workers for five months – Barnes took time out to photograph the local town, its people and surrounds. With more than a decade's practice in the art of photography (and with the help of much-improved equipment and processing), his landscapes are carefully composed, aesthetic and lovely works of photographic art.

Above and opposite, top row and middle row left: Prints from Henry Barnes' personal album. The unlabelled image is a Sydney street scene, possibly taken near the museum.
Photographer: Henry Barnes

Opposite, middle row right and bottom row: Prints from the 'Photographs by the Staff of the Australian Museum' album.
Photographer: Henry Barnes

Pages 118–19: Striped Marlin,
Kajikia audax.
Photographer: Henry Barnes

Middle Harbour looking South.

Residence of Mr. Sibbald – Wellington.

Bell River – Wellington.

Bell River near Wellington, N.S.W.

Falls – Jenolan Caves.

ARTISANS AND
TECHNICIANS

PART II

CHAPTER FOUR

THE AUSTRALIAN MUSEUM PHOTOGRAPHER, 1857–1893

PHOTOGRAPHY

UNDER this general name is distinguished one of the most beautiful discoveries of the present century; so remarkable for scientific advancement. Sun-painting ... pictures formed on the ground glass of the camera obscura, arrested and rendered permanent to our admiring gaze by a simple chemical process ... rendered as truly as the reflection from a mirror:– This is literally 'holding the mirror up to nature'.

Sydney Morning Herald, 10 May 1855

A VISUAL WITNESS TO MUSEUM SCIENCE AND PRACTICE

For the Australian Museum's early scientists, like Gerard Krefft and Edward Ramsay, photography promised enticing benefits. It acted as a credible empirical witness to new discoveries (for example, of animals, people, natural phenomena and field sites); it allowed scientists to share prints among colleagues far and wide, both at home and abroad; and it created accurate, high-resolution images that could be studied in detail, used for ongoing reference, and reproduced as illustrations in scientific publications.

Scattered throughout the museum's archive and recorded history is evidence of photography's growing prominence and usefulness. In the museum's Trust Minutes, the first reference to photography appeared in 1857; in 1868, a new camera lens was acquired; in 1873, £4 11s was spent on photographic chemicals. By then, photographic exchanges were commonplace and there were even occasional requests from external practitioners to use the museum's 'photographique atelier' (which was really just a shed outside the main museum building, shared with the taxidermy and articulation functions).[1] A purpose-built photographic studio was not constructed until 1897.

The glass plate photographic collection at the Australian Museum is a witness to the priorities, pursuits and practices of its scientists in the 19th century. The subject matter of the images tells us what was considered important enough to document and worth the time-consuming labour of photography. The physical glass plates allow us to track the progress of skills, technology and photographic resources over time. And despite many of the glass plates being more than 150 years old, the high level of resolution means they can be enlarged to reveal every detail of the image with astonishing, impeccable clarity.

Page 120: Gorgonian Coral or Sea Fan, *Junceella* sp. This large coral fan and stone were caught in a fisherman's line off Long Reef, near Manly in Sydney, in 1898. Shortly afterwards it was purchased by the museum. Zoologist Thomas Whitelegge photographed and washed the specimen, recovering more than 65 different marine species and at least 372 individual specimens from within the crevices of the stone. He published his photograph and species list in 1900, proclaiming the find as a wonderful demonstration of the extreme richness of the bottom fauna along Sydney's coastline.
Photographer: Thomas Whitelegge

Page 121: Nest and eggs of the Satin Bowerbird, *Ptilonorhynchus violaceus*.
Photographer: Henry Barnes, jnr

Opposite: A list of photographic chemicals required by the Australian Museum, written by Edward Ramsay in the Curator's Report dated 5 July 1881. Among the supplies Ramsay lists two types of collodion, pure ether, nitrate of silver, bromide of ammonia, pyrogallic acid, hyposulphite of soda, and one ream of albumenised paper.

…als & Glass Bottles

It is requested that all communications with the Museum be addressed to the Trustees.

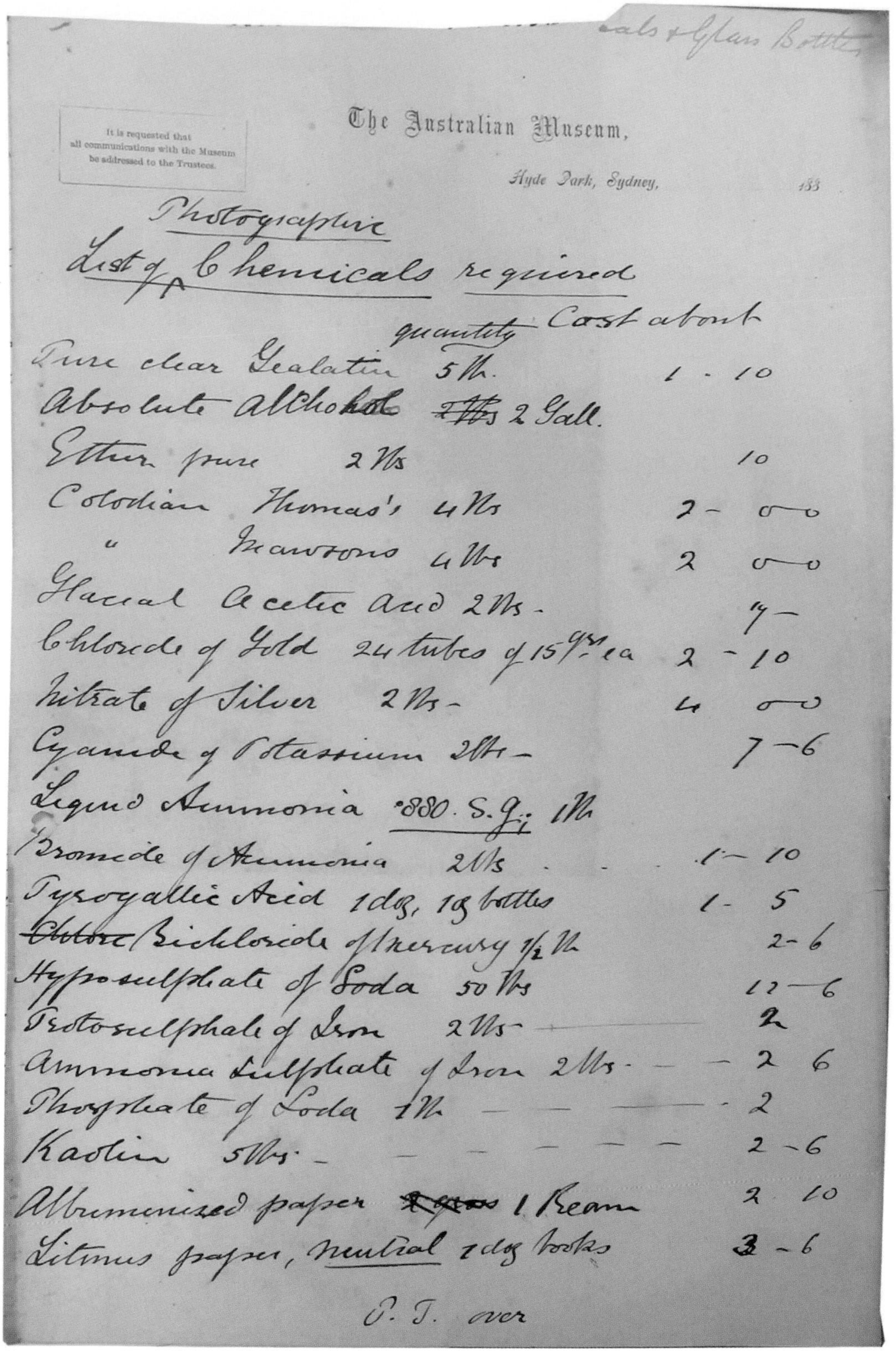

The Australian Museum,

Hyde Park, Sydney, 188

Photographic

List of Chemicals required

	quantity	Cost about
Pure clear Gealatin	5 lb.	1 - 10
Absolute Alchohol	~~2 lbs~~ 2 Gall.	
Ether pure	2 lbs	10
Colodian Thomas's	4 lbs	2 - 0 0
" Mawsons	4 lbs	2 0 0
Glacial Acetic Acid	2 lbs -	7 -
Chloride of Gold	24 tubes of 15 grs ea	2 - 10
Nitrate of Silver	2 lbs -	4 0 0
Cyanide of Potassium	2 lbs -	7 - 6
Liquid Ammonia	·880. S.G. 1 lb	
Bromide of Ammonia	2 lbs	1 - 10
Pyrogallic Acid	1 doz. 1 oz bottles	1 - 5
~~Chlori~~ Bichloride of Mercury	½ lb	2 - 6
Hyposulphate of Soda	50 lbs	12 - 6
Protosulphate of Iron	2 lbs	2
Ammonia Sulphate of Iron	2 lbs	2 6
Phosphate of Soda	1 lb	2
Kaolin	5 lbs	2 - 6
Albumenised paper	~~[illegible]~~ 1 Ream	2 10
Litmus paper, neutral	1 doz books	3 - 6

P. T. over

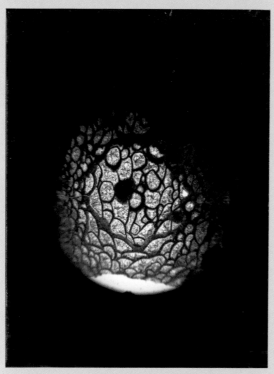
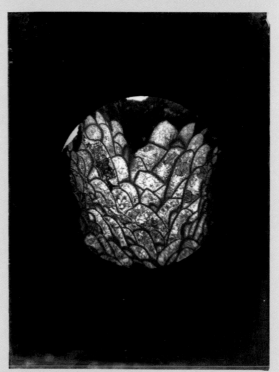

REVEALING SMALL WORLDS

PHOTOMICROGRAPHY

Experimentation with photomicrography began soon after the invention of photography in the 1830s. Microscopy was already well established as a scientific tool, but now a camera could be attached to the top of the microscope using a clamp or stand to create a 'double lens'. The magnified images produced were called photomicrographs. Permanently fixing true-to-life, microscopic detail, these images were regarded as more 'objective' and 'scientific' than the hand-drawn and engraved illustrations they replaced.

The first microscopic images were captured by English scientist Henry Fox Talbot in 1835, but needed to be exposed for up to fifteen minutes and could only be magnified 20 times. Although further progress was made in the 1860s, it was not until improvements in lens technology and artificial lighting in the late 19th century that photomicrography became a standard practice in many scientific disciplines. Life could now be viewed at cellular level, and photomicrographs captured photographic images of algae, diatoms, bacteria, tissue sections and human blood for the first time.

The beautiful images produced in the process meant photomicrography also entered Victorian popular culture, becoming a genteel hobby. The results of this leisure activity could also be useful for science: amateur American photomicrographer Wilson Bentley took 5000 images of snowflakes over the course of forty years from 1885 to support his theory that no two snowflakes are the same.[2]

Photomicrography came late to the Australian Museum, and its development was led by laboratory scientists and not by the museum's regular photographers. Fish curator Edgar Ravenswood (ER) Waite was one of the first to apply this technology, using photomicrography to capture the detail and beauty of these coral specimens for the museum's curator, Robert Etheridge, jnr, in the early 1890s.[3]

Opposite: Photomicrographs of fossil coral specimens, made by ER Waite for palaeontologist Robert Etheridge, jnr, curator of the Australian Museum.

THE COLONY ADOPTS PHOTOGRAPHY

It did not take long for photography to reach the southern hemisphere: the first rudimentary imported cameras and glass plates were available in Sydney from the beginning of the 1840s. The earliest recorded daguerreotype taken in Australia (an image of Bridge Street, Sydney) was produced by visiting naval captain Augustin Lucas in 1841, and the country's first commercial studio photographer, George Goodman, was already working in Sydney by the end of 1842. He was said to have taken an astounding 1200 daguerreotype portraits in his first two months of operation.[4] Pioneering photographic techniques required prolonged exposures. Goodman's subjects, for example, had to endure up to 90 seconds of immobility and sweltering heat on the roof of Sydney's Royal Hotel for their portrait sittings. As subjects had to stay completely still to avoid blurring the finished image, a head clamp to prevent movement was part of every daguerreotype portrait photographer's equipment.

In the first four decades of the new technology, almost all photography in Sydney was commercial, and the market was soon flooded with photographers offering their services to the public. In the decades between 1860 and 1900, a total of 450 photography studios operated in Sydney.[5]

In 1859, professional Sydney-based photographer William Blackwood introduced the carte de visite to Australia. These small calling cards, featuring the sitter's portrait mounted on cardboard,

were so popular that they were collected in albums for safekeeping and display. Worldwide, they were produced and swapped in their millions.

Studios made, advertised and sold printed images of everything from celebrities, actors and actresses, to heroes of the goldfields, local events, ships in the harbour, accident survivors, Australian Aboriginal people and romantic outback scenes.

By the time they began their experiments in the early 1860s, museum photographers Henry Barnes and Gerard Krefft were part of an industry.

A DEVELOPING SCIENCE

In its pioneering days, photography was referred to in scientific terms: 'photogenic drawing' and 'chemical changes produced by solar radiation'. By the 1850s, however, writing on the process was broader and more accessible. As new, more convenient techniques began to emerge, any interested amateur, with appropriate funds, could attempt photography without the benefit of a scientific background.

The first cameras were simple (a wooden box with glass plate holder and lens attached), but bulky and large. Because long exposures were required, photographers had to carry heavy wooden tripods to stabilise their images, along with the weighty glass plates and the chemicals and equipment needed for developing the plates.

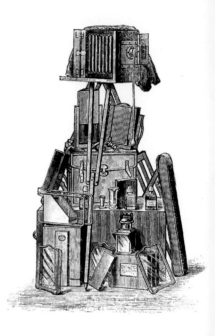

Above: *Marion's Practical Guide to Photography* by the Marion Photographic Company, 1885, advertised their complete photographic sets including glass plates and accessories.

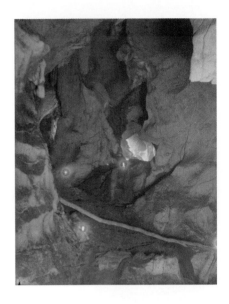  

The knowledge and skills required for the chemistry of photography were gleaned from information circulated from Europe and the United States. Manuals on photography were published from as early as 1839, and their authors (scientists, and later commercial photographers and stockists) were keen to claim the credit for certain formulations, and to satisfy the public obsession with the new medium. Accounts of the chemical processes involved were often published at length in newspapers and periodicals. The *Weekly Times* in Hobart, for example, published full chemical recipes for photographic sensitising, toning and fixing concoctions in its edition of 23 May 1863. Chemicals could also be purchased from pharmacists or direct from photographers' studios or, after the 1870s, from dedicated photographic supply shops. Still, the multi-step process of preparing glass plates, taking the photograph, developing the negative and then making a print was complicated, and equipment was expensive.

From 1881, however, the dry plate method made photography simpler (plates could now be stored and processed long after they were made), and the art immediately became more accessible. Popular access to and interest in photography did not become widespread until 1888,[6] when the first Kodak film camera was produced and became available in Australia. It was capable of taking 100 exposures, after which the owner returned the camera to the manufacturer for the film to be processed into prints. In 1896 a pocket version was released, followed in 1900 by the Kodak Brownie, the first mass-marketed camera.

Above: These three images taken in Wellington Caves, New South Wales, in 1869, display the variation in glass plates of different exposures. The first image, at left, is slightly overexposed; the second image has a good variation between shadows and highlights; and the final image is highly overexposed, rendering its details harder to discern. Photographer: Henry Barnes

CORAL MYSTERY

LETTUCE CORAL,
PECTINIA LACTUCA

The origins of this fragile beauty, the coral *Pectinia lactuca*, remain mysterious. The specimen was acquired before 1876, but was not registered when it first arrived at the museum – as was common in an institution suffering from growing and storage pains and a lack of scientific expertise. Most likely, the coral arrived as part of a large collection of material from Vanuatu, then known as the New Hebrides.

Many years later, in 1891, the coral was described by English marine biologist William Saville-Kent, who worked in Australia for some years and was later to become a famous reef photographer. At the time, Saville-Kent was working in Sydney, and he identified this 'magnificent specimen' with its 'great depths of calicinal valleys, their perpendicular walls and subeven non-lacinulate distal edges' as a new species, *Tridacophylla rectifolia*.[7]

That identification has since been overturned, and the coral – common in reef environments around the world – is now identified as *Pectinia lactuca*, or Lettuce Coral.

It is unclear when this series of photographs was taken, but the images were published in 1891 in *Records of the Australian Museum*, alongside Saville-Kent's description. The artistry of these photographs, and the use of close-ups and technical lighting to reveal the specimen's structure, suggest they may form one of Henry Barnes' more confident, later series.

It was another twenty years before the coral specimen received any more attention, and in 1907 it became a formal part of the collection, with registration number G7230. It is not surprising that by then the origin story of the coral had been lost or forgotten.

Opposite: Three views of Lettuce Coral, *Pectinia lactuca*, 1891. Photographer: Henry Barnes

THE GLASS PLATE TECHNIQUE

VANESSA LOW

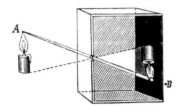

Photography in the 19th century was a highly specific art form that required skill, patience and an experimental spirit – but above all it was a time-consuming, complex and laborious chemical process. To capture the desired image, a photographer had to be diligent in their preparation and execution (particularly as some chemicals were toxic and unstable). One wrong step and the image would fail to appear, representing a considerable waste of time and effort.

The evolution of photography involved experiments with a variety of chemical processes to produce different outputs, such as daguerreotypes (a one-off image permanently fixed on a copper plate), tintypes (a permanent image on an enamel-coated sheet of iron) and calotypes (a reusable negative image produced on paper). However, the most widespread method – and the one used by the Australian Museum's photographers – was the wet plate (or wet collodion) process. This glass plate technique dominated the photography market from its introduction in 1851 until around 1881, when the much more convenient dry plate process arrived and swiftly superseded the wet collodion method.[8] The dry plate method employed many of the same techniques as its predecessor, but made the process achievable in a fraction of the time.

The aim of both the wet collodion and dry plate methods was to capture a silver image, bonded onto a sheet of glass (referred to as the glass plate).

In wet plate photography, collodion (a clear solution of cellulose nitrate in a mixture of ether and alcohol) was used to create a thin, translucent film on a piece of glass. It was then immersed in light-sensitive silver nitrate, which stuck to the film, creating a sensitised surface.

In the dry plate process, instead of creating the liquid chemical reaction themselves, photographers would buy the sensitised glass plates pre-prepared: silver chemicals were already bonded to the glass plate in a layer of gelatin. The dry plates had to be stored in darkness, but when the time came to take a photograph, they were ready to go without any additional preparation.

READYING THE CHEMICALS AND TOOLS

Before they even approached the camera, the photographer had to ensure that their chemicals and tools were all in order. If using the wet plate technique, they would need to treat the collodion by mixing it with iodising chemicals; this was done between a day and a week in advance. Later in the process, these chemicals would cause the all-important reaction with the silver nitrate, making it light sensitive. If using the dry process, the photographer would check their stocks of pre-sensitised dry plates. Quantities of developing and fixing chemicals needed to be accounted for, and tools (such as tongs or porcelain trays) cleaned.

Above: Due to the way light enters a pinhole, the image projected inside a camera would appear back to front and upside down. This could make framing a scene challenging for a photographer, as objects would appear opposite to how they looked in real life.

Above: Smallscale Bullseye,
***Pempheris compressa*, with**
a ruler to show scale.
Photographer: Henry Barnes

Above: A newly received whale in the museum's garden, photographed to show off its skull and baleen, with a ruler positioned to provide scale. The taxidermy shed can be seen in the background.
Photographer: Henry Barnes

Opposite: Lord Howe Butterflyfish, *Amphichaetodon howensis*. The tiny nails holding the fish in place are just visible in the photograph.
Photographer: Henry Barnes, jnr

SETTING THE SCENE

While it was common to photograph taxidermied animals for the museum's collection, occasionally the specimens in the images were freshly caught: as in the case of the still-glistening fish shown on page 131. Small specimens – teeth, for example, or certain fish species – would be set out on a blank surface, positioned next to a ruler for scale. Taxidermied mammals and birds might be arranged as part of a group (discussed in more detail in chapter five), in a carefully posed composition. Fossils or bones were often categorised and presented by type, size or shape. And larger skeletons and animals would be photographed outdoors, against a clean, studio-like backdrop.

Composing, styling and lighting the specimens usually required considerable time and effort – a spontaneous snap was almost impossible.

A variety of approaches to styling are evident across the early museum photographs. Sometimes the set-up seems rather perfunctory: a large white cloth thrown carelessly over a table to act as a backdrop; a whale haphazardly positioned on the grass with a ruler resting precariously atop its jaw. However, there are also very delicate and considered moments of styling: two small birds perched on a branch as if in conversation (see page 163); a scene of Superb Lyrebirds positioned among carefully arranged foliage (see page 168); and a Tasmanian Devil gripping its prey in its jaws while bounding through a grassy landscape in the museum garden (see page 136).

1936

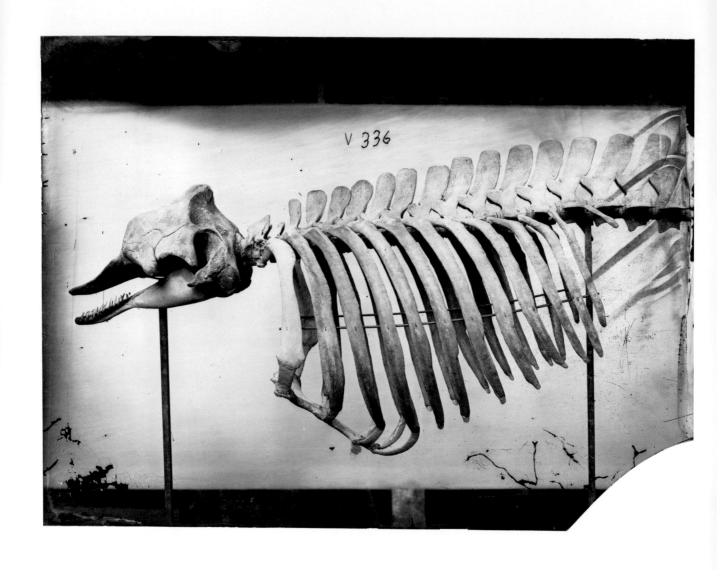

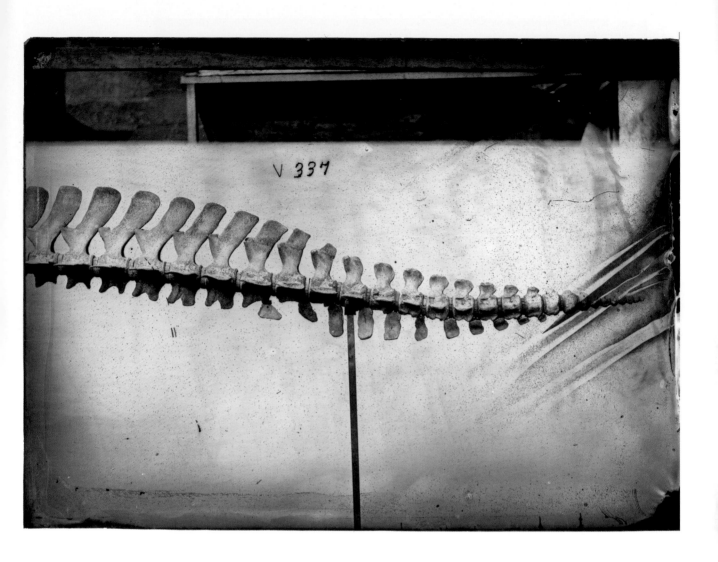

Above: Pygmy Sperm Whale, *Kogia breviceps*, collected at Manly beach in Sydney and photographed in two parts in the museum courtyard. This is possibly the type specimen for Krefft's *Euphysetes macelayi*, which he first described in 1866.
Photographer: Henry Barnes

Lighting was of the utmost importance for photography. Without the aid of artificial light (bar the use of candles, and oil or gas lamps), museum photographers sought locations that were lit evenly and well. Large objects were often photographed outside, bathed in daylight. Shadows were not always desirable, so care was taken to position specimens in ways that prevented shadows being cast. Smaller bones were sometimes photographed on glass shelves.

As their confidence increased, there are also instances of the photographers using more artistic or specialist lighting; for example, in an image of a backlit coral, its spiny fingers threading through the dense black background (see page 120).

Once the objects were in place, the photographer would frame the scene in the camera. They did this by looking through a rectangular piece of glass at the back of the apparatus (the 'ground glass'). There were no zoom lenses at this time (lenses were 'fixed'), so if the photographer wanted to adjust the framing, they would need to physically pick up and move the camera and tripod. To focus the image, they could shift the camera or adjust the lens (depending on what type they were using). This could be tricky. Because of the way the light entered the pinhole and was projected inside, the image would appear back to front and upside down on the ground glass.

Once all of this was done, the photographer would go to the darkroom to commence the chemical procedure.

Above: Taxidermied Tasmanian Devil, *Sarcophilus harrisii*, posed with prey in the museum's garden. Photographer: Henry Barnes

Opposite: Half-hatched Emu egg balanced on a glass test tube. Spots on the image indicate that the glass plate may not have been cleaned properly before use, or that the chemicals were contaminated. Photographer: Henry Barnes, jnr

Above, left to right: Small objects
such as this Southern Shovel-
nosed Cray, *Arctides antipodarum*,
were sometimes suspended by

Above: The white paint masking on this glass plate has been digitally removed and the image has been reversed to show the 'positive' image. The semi-dried Tiger Shark, *Galeocerdo rayneri*, was photographed on a table outside the taxidermy shed. Photographer: Henry Barnes

TRANSFORMING PLAIN GLASS INTO A PHOTOGRAPHIC PLATE

Whether obtained from photo studios or suppliers, or cut by the practitioners themselves, a glass plate needed to be carefully inspected and cleaned before use. Ammonia, alcohol, caustic potash (potassium hydroxide) or soda could be rubbed on the glass to clean it, and a camel's hair brush used for dusting.

If using the wet plate process, the photographer would coat the glass evenly with a layer of collodion. This was done by holding the plate flat in one hand and using the other hand to pour a pool of liquid collodion into the centre of the glass. The glass was then tilted from corner to corner so that the liquid flowed evenly across the plate, ideally touching each edge without spilling over it. This process, called 'flowing the plate', required both delicacy and efficiency, as the chemical would begin evaporating immediately. Streaks could appear on the final image if the collodion was not applied evenly; and if the photographer's fingers touched the chemically treated surface, oil or fat could be dissolved from the skin and cause marks on the picture.

Next, the photographer had to transform the plate into a light-sensitive object, a process sometimes referred to as 'exciting the plate'. Working in the darkroom, often under orange or red light, the photographer would insert the glass plate into a porcelain tray filled with a silver nitrate solution (which would react with the collodion to become light sensitive). The liquid in the tray needed to be about 6.5 millimetres deep in order to cover the plate without air bubbles forming, as these would cause spots on the final picture. The plate was immersed swiftly, with the collodion side facing downwards, and the porcelain bath gently lifted and rocked to ensure a smooth and rapid flow of the chemicals across the surface.

The whole process could take the photographer a few minutes, more or less, depending on the weather – it was slower in winter and faster in the heat. It took experience gained through experimentation to know when a plate looked just right.

Once ready, the glass plate was removed from the bath and placed upright on blotting paper to drain. Then, ensuring that the sensitised side was facing down, the photographer would carefully place the glass into a plate holder: a wooden frame, sealed against the light, that held the glass plate. Because the process involved liquid chemicals, the photographer had to act quickly; they had about ten minutes before the chemical balance was lost.

On the other hand, if a dry plate was being used, all of the above steps could be skipped and a pre-sensitised glass plate could be inserted straight into the plate holder – much more efficient (and without the time constraints)!

Polishing the plate.

Coating the plate.

Sensitising the plate.

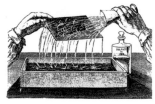

Developing the plate.

Above: Steps in the wet plate process. As photography was a relatively new industry, each photographer's equipment and procedure would have varied slightly.

LETTING THE LIGHT DRAW

With the glass secured in the plate holder, the photographer could now leave the darkroom and return to the scene they wanted to photograph. Checking the view in the ground glass for a final time, they would insert the plate holder into the camera and remove one of its securing flaps (the 'dark slide'). Finally, with an eye on a pocket watch, the photographer would remove the lens cap and expose the plate to light.

After the desired exposure time had passed (calculated depending on the time of day, availability of light and sensitivity of the chemicals), the lens cap would be replaced and the dark slide returned, once again securing the plate in darkness. The art of determining the right exposure would have been mastered over time. As English scientist Robert Hunt summarised in his 1857 *A Manual of Photography*:

> one photographer may prepare his plate in such a manner that even in a few seconds his negative picture is too dark, another may find that he cannot obtain a negative picture dark enough in several minutes. Experiment alone can determine for each individual the duration of exposure for his sensitive plate in the camera obscura.[9]

The photographer would then remove the plate holder and return to the darkroom.

Above: Nest of the Noisy Friarbird, *Philemon corniculatus*. Photographer: Henry Barnes

Opposite: Green Grocer Cicada, *Cyclochila australasiae* (top) and Double Drummer Cicada, *Thopha saccata* (bottom). The emulsion has cracked and peeled around the edges of the glass plate, demonstrating how thin and fragile the collodion layer was. Photographer: Henry Barnes, jnr

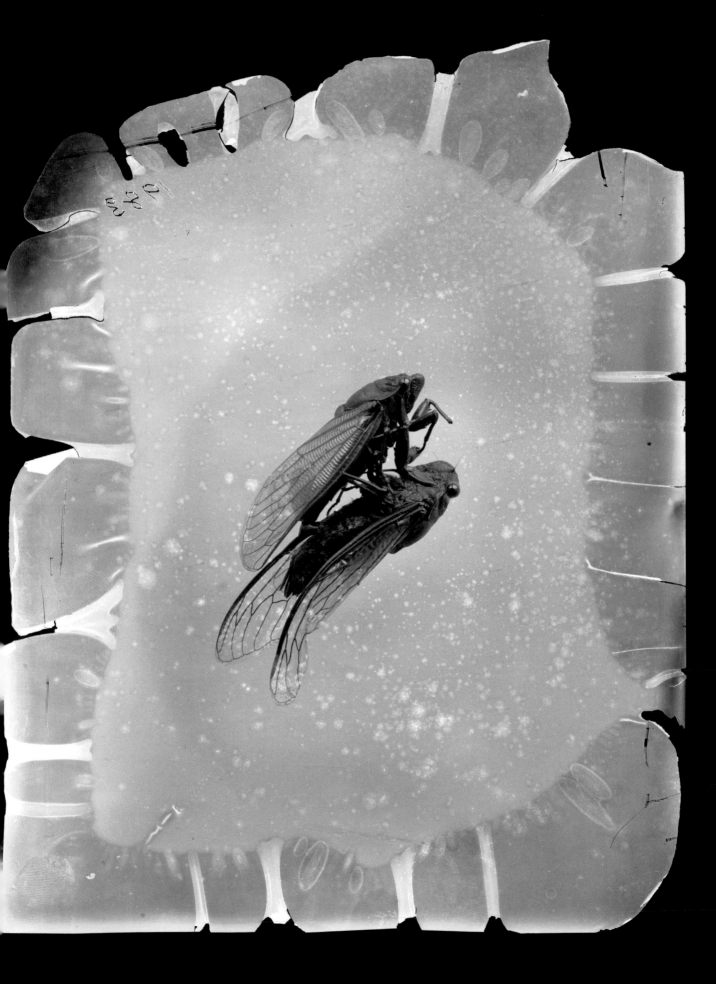

REVEALING THE IMAGE

Back in the darkroom, the plate holder was opened and the glass plate allowed to drop onto a flattened, waiting hand. No image would be apparent to the naked eye: although the image had formed chemically on the surface of the plate, it was still invisible at this stage. To provoke the image to appear, a developer solution would be poured across the plate in an even, gentle motion. The developer contained acid-reducing agents that caused light-exposed silver to darken, and the image would begin to appear, highlights first (light clothing, or faces, for example) and then the shadows, darkening over time. When satisfied with the image, the photographer would pour water across the glass to halt the chemical process.

The next step was to fix the image to avoid future darkening caused by light exposure. The Victorian-era photographer would have done this by washing the image with a liquid fixer, often containing sodium thiosulfate (at that time commonly known as 'hyposulphite of soda' or 'hypo') or cyanide. These chemicals ensured that all unused, light-sensitive silver particles were removed from the plate, thereby stabilising the image. The plate was then washed with water.

Finally, to protect the surface from becoming damaged or scratched, the plate was warmed and varnish poured carefully over the exposed side of the plate. It was then left to dry for a few hours.

Left: This photograph of the Australian Museum exterior has been developed unevenly – perhaps dipped into a developer bath on one side, then dipped on the other side, resulting in the dark 'overlapped' area in the centre of the image. The photograph is also 'foggy', and has spots and markings indicating that it has been affected by impure chemicals. Photographer: Henry Barnes

Opposite: Recently prepared whale skull, posed to show baleen. Photographer: Henry Barnes

PRINTING THE PHOTOGRAPH

The final result (for both wet and dry process glass plates) was a negative image. This meant that when the glass was exposed onto paper, light filtered through the surface and left an inverse image – a positive, lifelike print. Negative glass plates could be used again and again, making them ideal for a museum scientist wanting to share their discoveries or research.

The most popular way to print photographs in the mid- to late 19th century was on albumenised paper, made by floating paper in a bath of egg white. (Another name for egg white is albumen, hence the name 'albumenised paper'.) Albumen contains proteins called albumins, and was a readily available and effective bonding agent. So popular were albumen prints that an 1870 issue of Adelaide's *Evening Journal* remarked:

> The annual consumption of eggs in photography is nearly a million in the United States alone, while the number used on this side [of] the Atlantic is probably at least three or four times as great. Hence it may be estimated that not less than five million inchoate fowls are sacrificed every year in the production of photographic portraits.[10]

The albumenised paper, whether handmade or purchased, was then delicately immersed in a bath of silver nitrate solution to be sensitised. The paper would be left for one to three minutes and then hung up to dry.

To make the print, the glass plate was secured in a wooden frame (the 'printing frame') with the collodion side facing upwards. The albumenised paper was placed on the glass and the back of the frame closed to compress the two items together. The frame was then exposed to light – next to a window, or outdoors – and the photographer would watch as the image appeared. This may have taken anywhere from a few minutes to an hour, depending on the sensitivity of the paper and the availability of the light. When the image had developed to the photographer's liking, it was removed from the light.

The paper was then immersed in a flat dish containing fixer (most commonly a solution of sodium thiosulfate and water; not to be confused with the fixer used for making the glass plate) for approximately ten minutes. After this, it was washed thoroughly in a water bath to remove any excess chemicals. The photographer had to be absolutely certain that no chemicals remained on the paper, as this would cause the image to react to light, and gradually darken. To ensure the paper was chemical free, the water in the bath would be poured away and refilled up to six times – a laborious but essential step. The image was finally laid flat on clean blotting paper or hung up to dry.

Once a print was completed, the museum photographer could use it in any way they wished. This could include pasting the print into one of the museum's large categorised albums (dedicated, for example, to 'Fish' or 'Fossils'). As discussed in chapter three, these albums were used both for display and as a reference resource. They also preserved the image for posterity, safeguarding against the original glass plate being broken, damaged, destroyed or lost. Although the museum still holds a large number of the original glass plates in excellent condition, many prints in the albums are the only copies of a particular image – and the only remaining proof that certain glass plates ever existed.

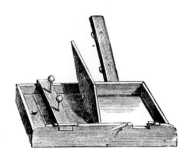

Above: Printing frames, which were made of wood, had opening doors on the back so that negatives and sensitised paper could be sandwiched inside to make clear, accurate contact prints.

Opposite, top and bottom: If an image was to be published as an illustration for reference in a scientific journal or book, the photographer would often want to show the specimen without the background. In this case, they would 'mask' the background by painting around it with white paint: when the image was printed, no light could pass through the paint, producing a 'clean' image showing only the subject. Photographer Henry Barnes

FIGHTING OVER TEETH

MARSUPIAL LION,
THYLACOLEO CARNIFEX

The first recorded fossil material of *Thylacoleo carnifex*, consisting of just a few isolated teeth, was recovered from Wellington Caves (discussed in chapter two) in New South Wales by Surveyor-General Major Thomas Mitchell in the 1830s. These few specimens were examined by palaeontologist Richard Owen at the Hunterian Museum in London – but without much knowledge of the context of this find, Owen was unable to offer an explanation for such puzzling remains.

The first *Thylacoleo* fossils to be formally identified by science came from Lake Colongulac in western Victoria. Richard Owen, by then at the British Museum, received these more revealing bones for study in 1855 and published a description of them in 1859, in which he named the creature *Thylacoleo carnifex*, meaning 'Marsupial Lion'. He interpreted the bones as the remains of a great marsupial carnivore, based on the carnassial-like premolars – teeth specially adapted for the shearing of flesh. He stated that 'it was one of the fellest and most destructive of predatory beasts'.[11]

However, Australian Museum trustee William Sharp Macleay disagreed and wrote to the *Sydney Morning Herald* the same year defending the maligned marsupial, declaring that it was 'A very gentle beast, and of good conscience'.[12] In 1866, in support of Macleay's view, Gerard Krefft published his opinion that *Thylacoleo* 'was not much more carnivorous than the Phalangers (possums) of present time'.[13]

Following further excavations at Wellington Caves later that year, and again in 1869, the discovery of more *Thylacoleo* body parts there – including an unusually large, clawed phalanx (thumb bone) – reignited the debate.

Seeing a photo of the phalanx, Richard Owen attributed it to a large carnivore and therefore to *Thylacoleo*. Krefft, however, had his doubts, and thought it could belong to some sort of Australian Giant Sloth. He and influential English naturalist William Henry Flower continued to think of *Thylacoleo* as a herbivore, based partly on its proximity in the archaeological record to other diprotodont marsupials, most of which were herbivores.

Richard Owen, on the other hand, was following French naturalist Georges Cuvier's principle that the first task in the study of a fossil animal was to explore the form of the molars to determine whether the animal was a carnivore or herbivore. He continued to argue that *Thylacoleo*'s general dentition resembled that of a carnivore and that the large, serrated upper incisors evidently had a laniary (tearing) function.

The argument regarding *Thylacoleo*'s diet and lifestyle, at this point, was based solely upon fossilised teeth and disarticulated body parts, the identification of which was partly circumstantial. It wasn't until 1966 that the first near-complete skeleton was found at Moree, northern New South Wales, and the picture became clearer.

Today *Thylacoleo carnifex* is identified as the largest carnivorous Australian mammal known. Owen's insistence on following the dentition was vindicated – but Krefft, Macleay and Flower weren't so far from the mark. Most palaeontologists still think *Thylacoleo*'s ancestors were herbivores, with the Marsupial Lion originating either from a possum ancestor (Phalangeroidea) or from a branch of the vombatiform line that includes herbivorous wombats and koalas.

Above and left: *Thylacoleo carnifex* is commonly known as the Marsupial Lion. It was first described by British palaeontologist Richard Owen. Photographer: Henry Barnes

CHAPTER FIVE

THE ART OF TAXIDERMY AND ARTICULATION

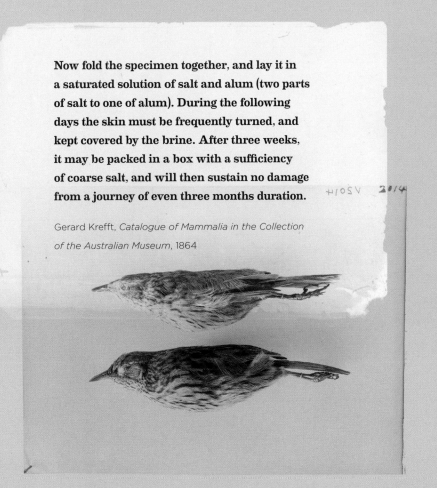

Now fold the specimen together, and lay it in a saturated solution of salt and alum (two parts of salt to one of alum). During the following days the skin must be frequently turned, and kept covered by the brine. After three weeks, it may be packed in a box with a sufficiency of coarse salt, and will then sustain no damage from a journey of even three months duration.

Gerard Krefft, *Catalogue of Mammalia in the Collection of the Australian Museum*, 1864

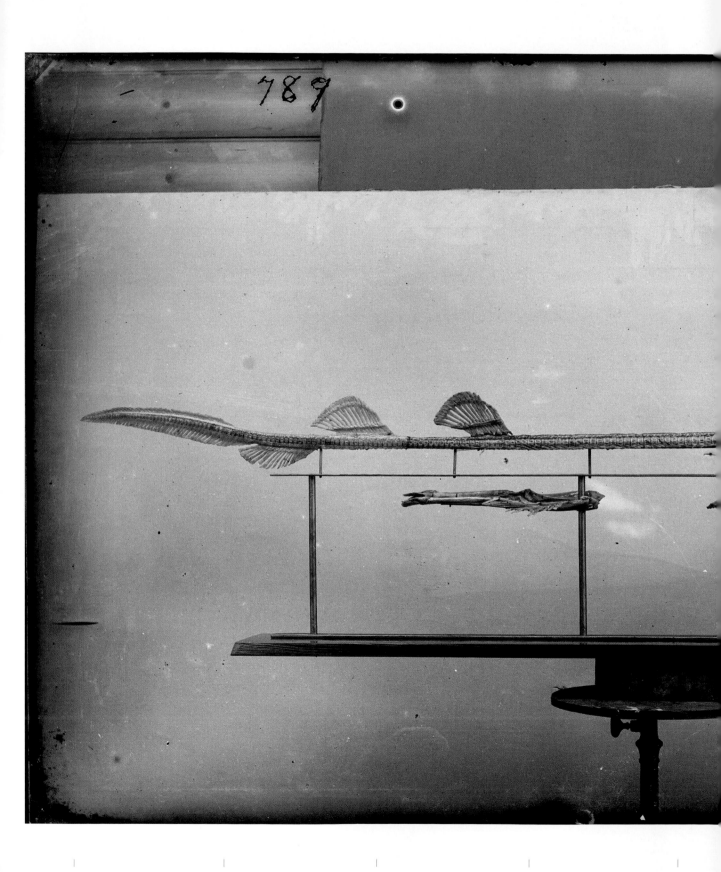

Like the early museum photographers, most 19th-century museum taxidermists and articulators learnt their skills on the job, either through experience gained in field collecting or in the workshop preparing animals for display. Unlike taxonomy (naming, describing and classifying animals into groups with other similar animals), which was increasingly reserved for specialised scientific workers, taxidermy was seen as an artisan trade, and was often passed down through family connections. At the Australian Museum, three members of the extended Barnes family worked as taxidermists – and as photographers.

Women could also make a living as taxidermists.[1] The Australian Museum's first professional female employee was taxidermist Jane Tost, who joined the institution in 1864. So highly valued were her skills that she was employed full time on the same salary as the male taxidermists, including her husband Charles.[2] In mid-19th-century Sydney, taxidermy was not just practised in the service of science: shop-front specimen dealers supplied natural history enthusiasts as well as the fashion and home furnishing and decorating markets.[3] By the 1850s there were no less than seven commercial taxidermy businesses in operation in the city. After she left the museum, Jane Tost set up her own business with her daughter Ada Rohu, and 'Tost and Rohu' traded in Sydney until the 1930s.

Taxidermy was the most sought-after and commercially useful skill – and therefore more common in the Sydney market – but articulation (setting up skeletons), casting and modelling were also very important parts of museum work and display, and became Henry Barnes' areas of special expertise.

Page 150: Skeleton of the Red Kangaroo, *Macropus rufus*. Photographer: Henry Barnes

Page 151: Type specimens of the Striated Fieldwren, *Calamanthus fuliginosus albiloris*, collected in 1865. Photographer: Henry Barnes, jnr

Left: Skeleton of the Spotted Wobbegong, *Orectolobus maculatus*. Photographer: Henry Barnes

THE GORILLA IN THE ROOM

GORILLA DISPLAYS AND THE EVOLUTIONARY DEBATE

There is no more potent symbol of the bitter Victorian-era fight over evolution than the gorilla. In 1859, French-American explorer Paul du Chaillu was the first modern European to come into contact with the animals, in Gabon, Central Africa. The unexpectedly human-like gorilla specimens that he shot and brought back to the West caused a public uproar when they went on display in London in 1860.

Commonly arranged in poses that served to emphasise their size and apparent menace, preserved and mounted gorilla specimens became a material display of the age's great scientific question; in effect, the 'gorilla wars' became a symbol of all that was at stake in debates over human origins and the role of evolution. In one corner was palaeontologist Richard Owen from the British Museum, defending human exceptionalism and the idea that every species was uniquely created and perfectly adapted. In the other was Charles Darwin, with the radical new idea that humans were descended from apes, supported by his vociferous ally Thomas Huxley.

When the gorillas arrived in London, Darwin's *On the Origin of Species* (1859) had only just been published, sparking controversy and wide debate. Owen (whose allies included many churchmen but also some leading scientists) had used differences in monkey and human brains to argue that they had been created separately and that it was impossible for apes to transmutate and become human. Darwin's supporter Huxley conducted his own monkey brain experiments, concluding that humans were not unique. At a famous debate in Oxford in 1860, Huxley scandalously declared that he was not ashamed to have a monkey as an ancestor.

Du Chaillu's sensational gorillas were displayed by Owen at the Royal Geographical Society in London in early 1860, and the public debate between Huxley and Owen continued in journals and scientific society meetings, attracting wide public attention and some satire. Thomas Huxley and the evolutionists eventually won, and Huxley's promotion of evolution shaped much of late-Victorian science.

AUSTRALIA MEETS THE GORILLAS

The first gorilla specimens in Australia were purchased by Gerard Krefft for the Australian Museum from London natural history dealer Edward Gerrard, arriving in Sydney in 1863. Unfortunately, the skeletons had been so haphazardly packaged for the long journey that they arrived in pieces, and were useless for research or display. Krefft was wretched with disappointment. He photographed the damaged lot and, mobilising his network of scientific contacts in London to amplify his complaints, wrote an angry letter to Richard Owen, enclosing a photograph of the mangled specimen in the opened case as evidence, while bemoaning the dealer's carelessness. 'Mr Gerrard did not even consider it worth his while (having received his money) to pack the smaller skeletons into separate cases.'[4]

Thus it fell to Frederick McCoy, director at the National Museum in Melbourne, to mount the first public display of gorilla specimens in Australia in 1865. A gift from the British Museum to the ardent anti-evolutionist McCoy, the specimen group included a large, 2 metre high male and a female with young, all shot by du Chaillu on his famous expedition. McCoy bragged that it was 'almost the perfection of taxidermy'. Even though their expressions might look human, he reassured readers, no-one should mistake

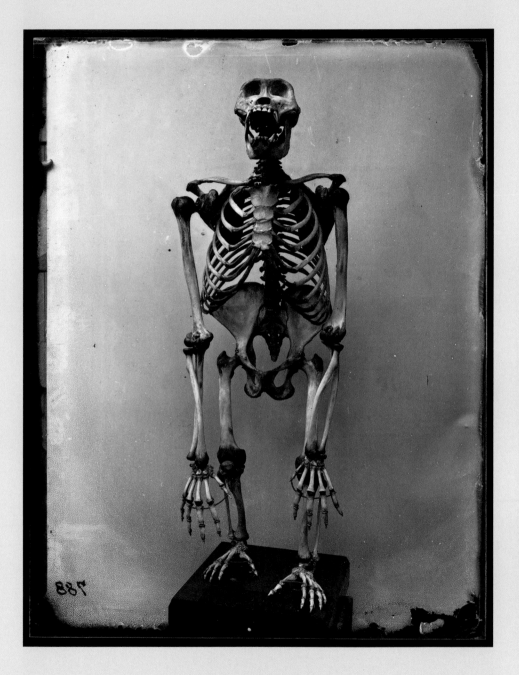

Right: Western Gorilla, *Gorilla gorilla*. While no longer suitable for display, the Australian Museum's gorillas are extremely valuable scientific resources. They provide a snapshot of the Western Gorilla population from Gabon as it existed nearly 140 years ago, one that can be used to monitor changes in that population over time.
Photographer: Henry Barnes

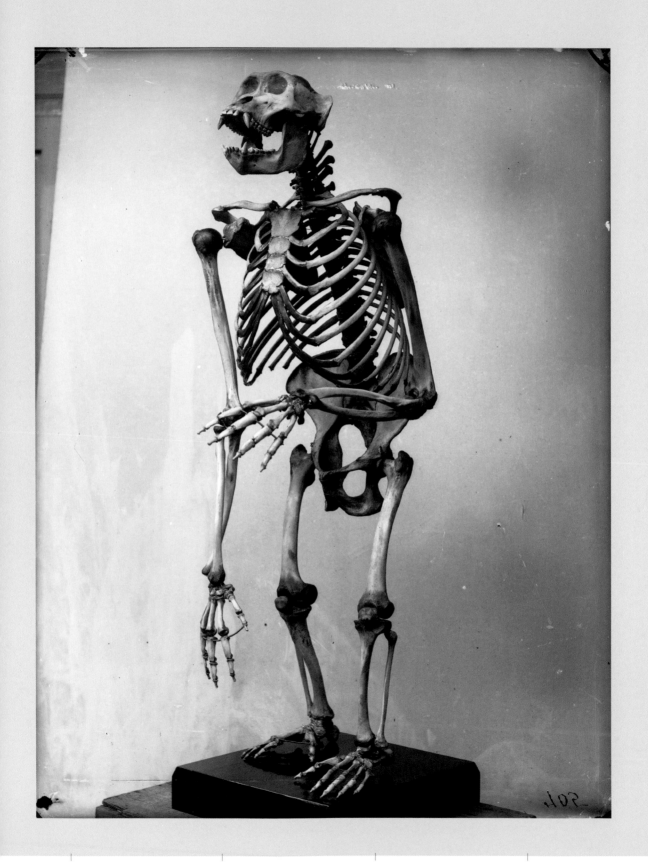

The Great Gorilla in the Sydney Museum.

that for a point of similarity or comparison, since above all the display showed 'how infinitely remote the creature is from humanity, and how monstrously writers have exaggerated the points of resemblance when endeavouring to show that man is only one phase of the gradual transmutation of animals'.[5]

In Sydney, photographs of gorillas had been part of the skeleton cases illustrating the 'five principal races of man' since 1868;[6] however, the first gorilla specimens did not go on display until 1885. Darwin's evolution revolution was still being debated, and placing a gorilla display on exhibit at the museum's front door was certainly provocative. The skins and skeletons again came from Gerrard, purchased by Edward Ramsay for £70. This time they arrived intact, although the skin was found to have been 'very badly preserved across the back. It required great care in handling, but the Taxidermist has made a fair job of it'.[7] Arranged in the preferred Victorian theatrical stance, the male gorilla is portrayed as aggressive and strong, posed at maximum height to emphasise its brute power and force and to provoke fear and awe in visitors (it was claimed that women in particular would 'pass them [the gorilla group] with visible signs and shudders of disgust').[8] Like the male's aggression, the compliant pose of the female and baby reflect Victorian gender stereotypes more than any deep biological knowledge of gorillas' habits and family relationships.

The Sydney gorilla group provoked a quick press reaction and seems to have promoted wonder, unease and disquiet in equal parts among visitors. Even if they were loath to accept its evolutionary narrative, some observers could not resist an urge to project the dark side of the human psyche

onto the gorilla.[9] One newspaper account reported that it was 'a horrible-looking brute, and not the less so from its hideous resemblance to man', and finished with this equivocal comment:

> Well, well my ugly friend, you may be descended from the same ancestor as myself, as some of the greatest scientists have laboured to demonstrate, but the proof must be established on a very irrefragable basis before I consent to accept you as the most distant of relatives.[10]

Twenty years later the gorillas were still on display at the museum, and still controversial, inspiring this 'Address to the Gorilla in Sydney Museum' by 'W.H.Y.', published in the *Sydney Mail and New South Wales Advertiser* in 1903:

> Helpless in strife,
> Shrieks thy victim for life!
> Hideous, grinning Gorilla, defying resistance.
> Insatiate, cruel blood-spiller,
> No beast in existence
> Could fill a
> Soul with a chiller
> Dread than thyself, the instiller
> Of fear in each bosom and thriller
> With horror; your form such a fright is,
> O, Troglodytes Gorilla![11]

With subsequent 20th-century discoveries about the complexity and nuance of gorilla behaviour, and the debunking of the myth of the gorilla 'alpha male' by naturalists George Schaller, Dian Fossey and others, Victorian gorillas like these are no longer deemed to be accurate representations and so are considered inappropriate for display at the Australian Museum. They were removed from exhibition in the 1950s.

TAXIDERMY

Such was the growth in the museum's collection after it opened its first gallery in 1857, and so great was the demand for specimens for display, that by 1867 there was a clear need for a separate taxidermy workshop. This was built in the museum's grounds, and it was also to become a makeshift photography studio. The following year, Henry Barnes was joined by a second permanent taxidermist, John A Thorpe, who remained at the museum for the next thirty-six years. In Thorpe's initial year of work at the institution, when he was probably still learning his craft, he and Barnes preserved the following new specimens, in addition to maintaining existing skeletons and mounts:

26 mammals
130 birds
12 reptiles and fishes
14 skeletons of various animals ...[12]

The taxidermists had an impressive workload and a continuous output. In the ten years from 1870, the two men prepared 4224 specimens, including 1480 bird skins, 1470 bird mounts, 92 mammal skins, 281 mammal mounts and 150 articulated skeletons.[13]

Taxidermy is the art of making a lifelike sculpture of an animal using its own skin. This is a three-part process, involving preservation of a dead animal's skin, followed by insertion of a frame or wires into the body cavity, before the form is finally stuffed and sewn together. Although manuals and pamphlets were available for Henry Barnes, John A Thorpe, Jane Tost and the other museum taxidermists to consult, these practical, inventive people also learnt from each other and sought to solve problems as they emerged. In 19th-century Australia, collectors and taxidermists faced particular challenges related to distance, climate and the country's unique animal forms. For those men and women working in the museum setting, flexibility and ingenuity were essential.

Barnes did not document his photographic work, and there are no photographs of the cameras, equipment or processes that he used. Nor did he document his taxidermy methods and techniques. As a result, the best evidence we have of the artistry and skill required of taxidermists – and of the limitations of their knowledge of chemistry, materials and techniques – are the museum's many surviving 19th-century taxidermied mounts.

Opposite: Holotype of the Short-beaked Echidna, *Tachyglossus aculeatus*, collected by John A Thorpe in Cape York, Queensland in 1869. The mounted specimen is still held in the museum's collections. Photographer: Henry Barnes

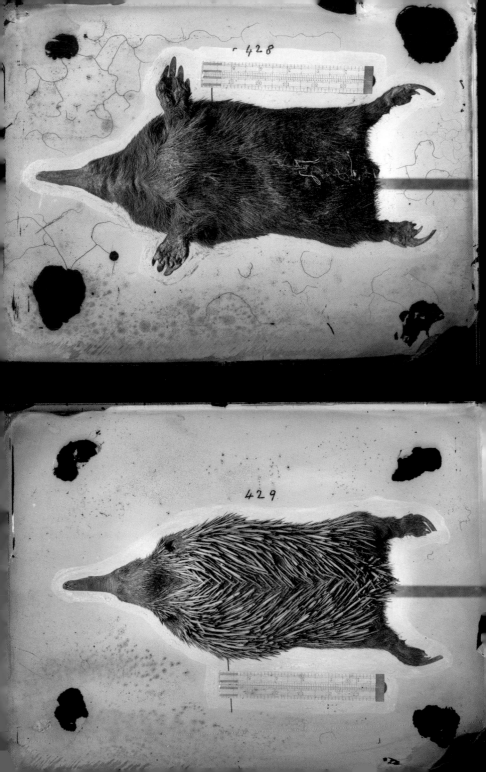

IN THE FIELD

The most important element for the museum's taxidermists in carrying out their work was a well-preserved specimen skin. For this, they were utterly reliant on the care and skill of field collectors.

Curators Gerard Krefft and Edward Ramsay were well aware of both the need for proper field treatment of specimens and the inconvenience and wastage that resulted when incorrectly prepared items arrived at the museum. Both published their own 'instructions for collectors', Krefft in 1860 and Ramsay in 1887.[14] Ramsay's short, portable pamphlet, *Hints for the Preservation of Specimens of Natural History*, was packed with tips to help the collector with the goal of delivering museum-quality skins, as well as whole specimens, which would not only remain intact but also retain the shape and likeness of the live animal.

An even better option for the museum was to send its own trained collectors into the field. As discussed in chapter three, Krefft despatched George Masters on numerous well-equipped collecting trips, and he returned with many thousands of specimens. The most famous of Masters' specimens are the collection of 26 Thylacines (*Thylacinus cynocephalus*) that he brought back from his Tasmanian field work in 1866 and 1867. Most of these specimens are still held in the museum's mammology collection as skins, mounts, skulls, bones and full skeletons, and are highly valued for research on this now-extinct animal. The museum also holds one precious and poignant preserved Thylacine pup, stored in alcohol in a large glass jar.

Before beginning the work of field preservation, collectors were instructed to record details of the animal's colour and shape. When documenting birds, they noted the orientation of the eyes, any fleshy parts and the animal's sex – since information about these essential features would be lost during treatment of the specimen. Ramsay suggested that for birds the date, locality and stomach contents should also be recorded. Drawings, notes and coloured illustrations would all help with later reconstructions.

Close attention was paid to the careful removal of bones, preferably through a single slit in the skin, with the skull and some major bones being retained. Following this, the drying and gentle stuffing of the skin was undertaken with great care, although a certain amount of shrinkage and distortion was inevitable. Some animal skins are very thin and delicate, particularly those of birds with soft or loose feathers (as well as those of fish, sharks and rays). Aware of the unique Australian conditions, Ramsay provided a special tip for bird collectors:

> if the weather be hot and the bird cannot be skinned at once, brush the bill, eyelids ... inside of the mouth, and other fleshy parts, as well as the abdomen under the feathers, and the vent, with the solution of mercuric chloride and camphor ... this will prevent those parts being blown by flies, and, to a great extent, arrest putrefaction.[15]

Opposite: No organs remain in a bird skin, including sex organs which would show whether it was male or female. The way the bird's legs are crossed (right over left or left over right) therefore indicates the sex of the specimen. Both these birds are adult females: Spinifex Bird, *Poodytes carteri* (left) and Broad-tailed Grassbird, *Schoenicola platyura* (right). Photographer: Henry Barnes, jnr

Stuffing material was chosen for its absorbency: moss, cotton, sawdust, straw and jute were all used. Ramsay also recommended that the body be stuffed with crushed, fresh, dry charcoal or wood ash 'to which may be added with advantage *grated or bruised* "galls", the excrescences found on the twigs and leaves of many of the gum trees (*Eucalypt*)'.[16]

Transporting large treated specimens was costly and difficult, and as early as 1864 the Australian Museum was offering to reimburse 'any reasonable charges for carriage or freight'.[17] Small animals and birds could be carefully pushed into a cylinder or preserved in 'strong spirits of wine', whereas the majority of skins were packed flat. Once received by the institution, those specimens chosen to become study skins for scientific research were left lying flat, as this made storage and examination much easier.

BRINGING A SPECIMEN TO LIFE

Back at the museum, the challenge for taxidermists was to transform tanned, macerated, pickled or dried skins and bones into lifelike animals for display. Many specimens were animals they may never have seen before, and certainly not in the wild. Instead, the taxidermists had to rely on collectors' field notes or illustrations in books, often created by people whose main reference was their experience of European animals. It is no wonder that early mounted quolls sometimes resemble ferrets! As more

specimens were collected and studied, and more accurate descriptions and illustrations were published in scientific papers and guides, taxidermy techniques improved and mounted specimens became more refined and truer to nature.

The basic tools required by a taxidermist were generic: scalpels, scissors, forceps, pliers, files, drills, chalk (for absorbing blood) and brushes, along with needles and thread. Australian Museum financial records from Krefft's era show that most supplies could be purchased locally in Sydney. And where supplies were not available, staff prepared the material themselves. For example, until galvanised wire became readily available in the late 19th century, iron wire had to be annealed (softened) in a flame before it could be bent to form the correctly shaped armature to support an animal's treated skin. Glue could also be specially mixed, rather than purchased. English naturalist William Swainson's recipe for glue was as follows: 'mix half a pound of gum Arabic with two ounces of white sugar candy; straining through a linen or horse-hair sieve. When it has become liquid, add a spoonful of starch or hair powder'. According to Swainson, this glue 'never spoils'.[18]

While some taxidermists working outside the museum made artificial eyes for their specimens, the Australian Museum favoured purchasing wholesale quantities of the realistic glass eyes that were commercially manufactured for both people and dolls.

Opposite: Green Catbirds, *Ailuroedus crassirostris* (a species of bowerbird), with their nest. Photographer: Henry Barnes

Pages 164–65: Eastern Quoll, *Dasyurus viverrinus*. This group of quolls was part of the museum's 'carnivorous marsupials' display in the Australian Hall, which included these Eastern Quolls (known then as the Native Cat of New South Wales), and other quolls, Tasmanian Devils, Thylacines, antechinus and 'Anteaters' (echidnas). Photographer: Henry Barnes

The volume of the museum's taxidermy operations – and the understanding that money was well spent on obtaining accurate, lifelike eyes – is evident in the detailed purchase order for £6 for glass eyes from supplier OJ Janson in 1894:

1½ gross [216] Clear Flints –
    size of a shilling
1½ gross [216] ditto – size of
    a sixpence (none smaller)
½ gross [72] dark brown for
    Kangaroos – size of a shilling
½ gross [72] violet for Satin Birds
2 gross [288] Nos 1 to 4
    chiefly browns and reds
2 gross [288] Nos 5 to 8
    all browns and reds
(no white or yellow eyes required).[19]

When it came to the look of the final specimen, faces were one area where the taxidermist could inject 'character' into an animal: through the positioning of the eyes and ears, by colouring the face, or by 'twisting the nose to depict curiosity; pushing in plaster to have teeth express ferocity; or tweaking a mouth to suggest watchfulness'.[20]

Of course, there are pitfalls to this sort of intervention. Nineteenth-century taxidermied Thylacine specimens, for example, are generally portrayed as aggressive, reflecting the general attitude towards this now-extinct species: which was that they were vermin and should be exterminated. Continuing conflicts between science and entertainment have often been manifested in arguments over taxidermy. In 1894, *Natural Science: A Monthly Review of Scientific Progress* ruled that 'The scene painter must not interfere with the scientist. A Museum is a place of truth before it's a place of art'.

Basic taxidermy techniques remained the same at the Australian Museum across the 19th century, as X-rays of remaining mammal and bird specimens with their primitive wire armature have shown. Although it continued to prepare Australasian specimens in house, by the late 19th century the museum was also purchasing exotics (African, American and some Asian specimens) as display mounts from catalogues and from suppliers like the famous Ward's Natural Science Establishment in Rochester, New York. Commercial suppliers were by this time producing 'sculpture-taxidermy' – where the body is first modelled in clay and then cast, before the skin and eyes are added. This became the norm for mammal specimens, with casting used for reptiles, fish and amphibians. By the mid-20th century, classical taxidermy of the stuffing and stitching sort practised by Barnes at the Australian Museum had disappeared from modern museums, overtaken by newer, more accurate, flexible and lifelike techniques.

ARSENIC

The history of taxidermy is also a history of chemicals and their uses. A variety of toxic chemicals have been employed in the quest to preserve skins, prevent their decay and ward off bug infestations.

Early taxidermists experimented with herb and spice recipes, often using cinnamon or tobacco to mask the smell

Above: Platypus group, *Ornithorhynchus anatinus*. Photographer: Henry Barnes

of rotting skins. None achieved long-term success, which had to wait until French apothecary Jean-Baptiste Bécoeur's invention and testing of arsenic soap in the mid-1700s. The preparation was spread on the inside of animal skins to prevent both biodeterioration and insect attacks.[21] Bécoeur kept the recipe a secret during his lifetime, but when the list of ingredients (camphor, arsenic oxide, carbonate of potash [potassium carbonate], soap and lime powder) was published thirty years after his death, the news spread quickly to England and across the world. It was eagerly adopted by taxidermists who were keen to be part of the booming – and lucrative – trade in natural history specimens.[22]

Writing in his 'Hints for the Preservation of Specimens of Natural History', Ramsay provided his own 'Receipt' [recipe] for Arsenic Soap:

> White Arsenic – 1 lb
> Common or Hard White Soap – 1 lb
> Salts of Tartar – 4 oz
> Lime in Powder – 3 oz
> Camphor – 2 oz

The ingredients were cooked and cooled, before being cut into cakes for storage and transport. The cakes could be softened with a brush and water when needed. Arsenic soap's simplicity, portability and effectiveness made it an essential part of all serious 19th-century naturalists' kit; in fact, few specimens prepared before its invention have survived.

Arsenic soap was only one of a number of approaches to specimen preservation in the 19th century. Thicker skins could be cleaned, then tanned either with salt, alum and water or with sulfuric acid, water and salt. Managing the preservation of the skin and its effects on the specimen was always a challenge – treatments often made skins vulnerable to shrinkage and distortion, especially if they were later rehydrated for posing and mounting.

Arsenic has not been widely used in museums since the early 1950s, replaced by less toxic borax. However, arsenic and other chemical residues remain in many of the historic specimens held in natural history museums, and are an ongoing problem for conservators tasked with the care and preservation of old mounts. Personal protection equipment must be used when handling specimens, and staff must avoid touching the internal areas of treated specimens unless wearing nitrile gloves.

Importantly, however, these often-degraded specimens may represent rare, endangered or extinct species and hold critical information such as species size, diet and regional variation. They remain an essential research tool for contemporary biodiversity and taxonomical research.

The same toxic chemicals that make handling so critical may also prove a boon for contemporary research into the history of taxidermy. Work has begun at the Australian Museum and the University of Sydney using analytical spectroscopy techniques to identify the chemical 'signatures' of particular taxidermists. If this is successful, it will help date and determine the provenance of specimens, and provide clues to the extent of different taxidermists' work across collections. It will also illuminate the Australian Museum's international trade and exchange with major early specimen exchange partners such as the Natural History Museum in London.[23]

THE THEATRE OF TAXIDERMY
ANIMAL GROUPS

Modern taxidermy dates from the era of the great International Exhibitions of the late 19th century. Starting with the 1851 Great Exhibition in London, they showcased science, technology and the wider world to the public – on a grand scale. At the Paris Exposition Universelle of 1855, displays for the first time included not just single taxidermied animals, but also groups of animals staged in their 'natural' habitats. These (slightly) more lifelike and theatrical groups were extremely popular with the public. They became a staple of International Exhibitions and animal-based entertainments, and were also a feature of commercial, ornamental taxidermy for domestic display.[24]

With demand growing for museums to reflect popular taste, animal groups began to appear in museum displays. The groups were first constructed and installed at the Australian Museum during the late 1880s, when gallery space was expanding and crowded display cases were being decluttered. Research specimens were no longer on show; instead, they were kept in safe storage, available to be accessed and investigated by scientists working behind the scenes. At the same time, the display purpose of habitat groups was shifting from merely entertaining visitors to teaching them about wildlife in the context of the animals' local environments. This marks a radical shift in the idea of the museum, since it was now the method of display or interpretation, rather than the specimens themselves, that was of greatest value.

Opposite: Superb Lyrebird, *Menura novaehollandiae*. Birds were some of the most popular museum exhibits, and by 1890 an incredible 700 of the then known 750 species were on display at the Australian Museum. These lyrebirds were part of a larger lyrebird display made in 1883, which included a family group, a nest and these two adult birds. The 1883 *Guidebook* describes 'the rich and varied notes of the Lyre Bird, far excelling those of the Song-thrush, and having immense powers of mimicry and ventriloquism'. Photographer: Henry Barnes

Around the world, museums took on the combined role of educating and entertaining the public, and adopted the new display techniques – which required preparators to be proficient in sculpture as well as taxidermy. American taxidermist and entrepreneur Henry Ward, who ran a taxidermy establishment in Rochester, New York, became world famous for his large, theatrically posed museum groups, which were available through mail-order catalogues and exported internationally. The Australian Museum purchased its first Ward exhibit specimens in the 1880s, and Ward became a major supplier of larger animals to the institution. Animal displays in museums worldwide continued to become ever more elaborate, and the art reached its apotheosis in the 20th-century natural history diorama. The first full-scale diorama was installed at the Australian Museum in 1923.

REPRODUCING NATURE

When it came to fast-moving, unpredictable, living creatures in the natural world, early glass plate cameras with their simple lenses (discussed in chapter four) were simply too slow to capture satisfactory images. In order to move into animal photography, canny commercial photographers drew on two resources: their experience with the props and artificial backgrounds used to simulate a broader context for posed human studio portraits; and the availability of taxidermied animal display specimens. While head clamps were required to keep human subjects still during long exposures, taxidermied animals were already immobile and camera ready. It appears that Sydney-based commercial photographer Charles Kerry borrowed some of the museum's animal groups during the 1890s to make postcards of Australian wildlife. Photographs of staged taxidermied tableaux like Kerry's provided the public for the first time with images of three-dimensional, live-looking animals in imagined versions of their natural habitats.

Opposite: Koala, *Phascolarctos cinereus*. This koala mother and baby were on display in a case in the museum's 'Australian Hall'. This is the description from the 1890 *Guidebook*: 'The Koala or Native Bear. These are quiet, peaceable animals; their fur is very thick and soft, generally grey in colour. Their favourite food is the leaves of the Eucalypti, though they eat some kinds of roots and in confinement will live a short time on bread and vegetables. The young is carried on the mother's back till it is able to climb alone'.
Photographer: Henry Barnes

ARTICULATION

Late-19th-century natural science was fascinated by bones. Whether the interest came from geologists looking for dinosaur remains or from evolutionists and their opponents, skeletons and the study of comparative anatomy revealed new truths about animals' forms, functions and behaviour, and bolstered arguments for and against their evolutionary adaptations.

The technique behind these skeleton displays is articulation, the process of cleaning and assembling an animal skeleton. It requires a mix of art, science and engineering – a perfect match for the practical skillset of the museum's hardworking taxidermists.

At the museum in the 1880s, bone and skeleton cases filled two Osteological Galleries. Arranged, as Krefft explained, 'with a view to illustrate the Homologies of the Vertebrata in the manner best adapted to meet the requirements of students in Comparative Anatomy',[25] their orderly ranks displayed their underlying unity of form, along with the Victorian view of the order of nature.

DISPLAYING SKELETONS

There were cases for mammals – carnivorous, ungulates [hoofed animals] and Australian cases – for birds, and for cartilaginous fishes such as rays and sharks. More cases highlighted the museum's expertise in excavating extinct Australian megafauna, with *Diprotodon*, *Nototherium* and *Thylacoleo* bones all on display. Hippopotamus, rhinoceros, crocodile, camel, giraffe and elephant skeletons filled the space. Whales and seals hung from the ceiling and, in the brutally racist 19th-century fashion, human skeletons and skulls were displayed alongside gorillas, chimpanzees and other monkeys.

The collection and display of human remains in museums was common in the 19th century and is part of the Australian Museum's inescapably colonialist history. The museum was founded with ideas of racial and cultural hierarchy that are now abhorrent but which have left a painful physical and ethical legacy in the museum's collections of human remains.

TRANSFORMING PRACTICE

Through the 20th century the museum began to acknowledge the consequences and legacy of its actions, and it has transformed its relationship to cultural collections and the rights of First Nations communities to self-representation, including decision-making over their collections in the museum's care. Since 1974, the museum has been actively supporting the return of Aboriginal and Torres Strait Islander secret sacred objects and human remains to their communities of origin through repatriation programs like the federal 'Return of Indigenous Cultural Property' protocols and funding. However, many of these remains were obtained unethically, often intersecting with ruthless racism and frontier violence. The long process of identifying individuals and negotiating for the return of remains to country

SKELETAL ARCHITECTURE

Most field collectors only dealt in animal skins (or sometimes preserved animals in spirits), since whole animals were difficult to prepare in the field and flattened skins were easier to transport and store. However, skeletons were popular with visitors and researchers alike, and much sought after for display. As a result, while articulation was time-consuming and delicate work, it was in high demand. Whale skeletons in particular were guaranteed crowd-pleasers. In 1849, a skeleton belonging to a Sperm Whale that had been hauled into Sydney Harbour was put on display (in a shelter) on the College Street side of the building, which was still under construction and not yet open to the public. The skeleton caused a public sensation, and a fence had to be built to hold back the crowd. The specimen eventually became part of the Long Gallery displays. Whale skeletons remain popular today. The huge Sperm Whale skeleton you can still see suspended high in the 'Wild Planet' Gallery has been on continuous display since it arrived at the museum in 1879.

The basic techniques of articulation also remain largely unchanged. The first task of the 19th-century articulator was to clean the bones, in order to remove all remaining flesh, connective tissue and ligaments. Animals were immersed in buckets or special 'maceration' tanks full of liquid (usually water). Sometimes this process would be supercharged with help from the larvae of flesh-eating insects.[26] In the 1870s, William John Macleay's taxidermist Edward Spalding 'was fortunate to have had access to abundant specialist "cleaners": specimens were sunk in open containers into the water of Elizabeth Bay to let sea life remove all edible parts'.[27]

Small animals could be treated whole. A whale, however, could be quite a challenge, with many of its huge bones too large to fit inside any tank. Curator William Sheridan Wall's 1849 attempt to obtain the magnificent specimen of a Sperm Whale for the museum's collection was much noted and documented at the time because of the 'disagreeable effluvium arising from it'. It took four days to flense (remove the flesh from) the animal, and Wall had great difficulty getting staff to undertake the task in the summer heat of December. As an alternative, stranded whales were (and still are) often buried in situ on beaches and recovered years later, once the flesh had entirely rotted away.

In the final stage of the cleaning process for all animals, bones were rinsed in clean water. Once dried, the bones could be bleached by exposure to sunlight to achieve the shiny white surface 'considered a great beauty in skeletons'.[28] The articulator would now have a pile of pristine bones to work with – and possibly no road map for their reconstruction.

Opposite, top: Blainville's Beaked Whale (also known as the Dense-beaked Whale), *Mesoplodon densirostris*. This was the first full skeleton ever recorded and the third-ever specimen in a collection anywhere in the world. The disarticulated skeleton is still held in the mammals collection. Photographer: Henry Barnes

Opposite, bottom: Skeleton of the Common Dolphin, *Delphinus delphis*. Photographer: Henry Barnes

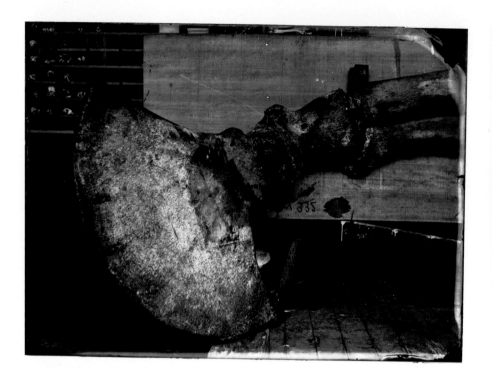
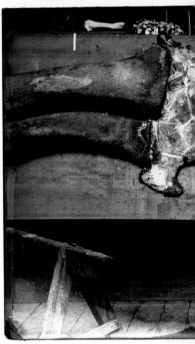

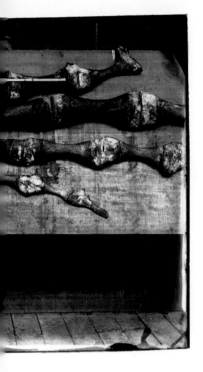

Above: This giant left flipper of a Humpback Whale, *Megaptera novaeangliae*, was collected at Little Bay in Sydney in the early 1870s. It was articulated immediately in the museum's new articulation shed and photographed in all its various parts and as a completed specimen (it also appears in the image on the title pages of this book). The flipper was featured in the early Osteological Galleries and is still held in the museum's collections.
Photographer: Henry Barnes

Hopefully, however, the bones had been kept in some order, and the articulator could then begin the painstaking task of reconstructing the animal's interior architecture. Holding the loose bones together – with elaborate supporting networks of wires, rods, nails, screws and glue – required ingenuity as well as skill. Henry Barnes was in charge of most of this work. In 1876, Ramsay supported the purchase of:

> [a] drilling machine by Barnes for the sum of £5 [as it] is invaluable when setting up skeletons, and takes drills from the size of the finest needle to that of ¼ inch, it has the advantages of being portable, easily worked and from the flexible arm, a hole can be drilled in any direction.[29]

Where bones were missing or damaged, infill plaster was used, based on best-guess estimates of the size and shape of the absent bones. Articulators could be inaccurate in their assumptions – or only as good as their own knowledge and access to published resources. Mistakes in shape and form could compound and multiply, as they did in 1871 when Barnes made the restored beak of a rare whale too large. Krefft sent a photograph of the 'beautifully prepared skeleton'[30] on to whale specialist Dr John Gray at the British Museum for identification and description. Gray noted that 'the form of the lower jaw gives a very peculiar

appearance to the skeleton' and went on to misidentify the whale as *Dioplodon sechellensis*. As Krefft later explained:

> It took Henry Barnes several days to patch the lower jaw together, and the beak being restored rather larger than the original (as was found out afterwards from Van Beneden and Gervais' splendid book); this misrepresentation multiplied by photographs sent to scientific institutions abroad, has caused considerable confusion ever since.[31]

Storing large skeletons has always been a problem for museums, and specimens are sometimes 'disarticulated' to save space. In 2008, a museum disarticulation project involving heritage skeletons revealed more information on how the taxidermists of the 1870s approached their work. While disarticulating a Strap-toothed Beaked Whale, *Mesoplodon layardii*, it was noted that hand-forged nails were not only driven through bones but also glued to ensure stability; other nails and wooden blocks had been used as wedges; there was widespread use of glue to hold bones together; and rods had been inserted through the spine. Leaving us another cryptic clue, the articulator (most likely Henry Barnes himself) had hidden in the depths of the cranial cavity scraps of local newspapers from 1869 and 1871.

Opposite: Greater Flamingo, *Phoenicopterus roseus*. This flamingo skeleton was presented to the museum by the trustees of the Zoological Society of New South Wales (Taronga Zoo) in 1893. The zoo helpfully donated many bird and animal specimens that year, when museum funds were halved because of a severe economic depression in New South Wales; as a result, opportunities for obtaining specimens were extremely limited. Photographer: Henry Barnes

THE PRINCE AND THE MANTA RAY

ALFRED MANTA,
MANTA ALFREDI

A PRINCELY VISIT

In the summer of 1868, Sydney was in the grip of royal fever. Young Prince Alfred, Duke of Edinburgh, the second son of Queen Victoria, was on the first-ever royal visit to the colony. In a six-month sweep, the prince had visited Adelaide, Melbourne, Tasmania and Queensland. The last stop on his tour was Sydney. Throughout his visit the prince had been greeted by large, enthusiastic, flag-waving crowds and carefully curated, elaborate public spectacles presented in his honour. In Adelaide, immense portraits of the prince adorned civic buildings, and 40 000 gas lights illuminated its public offices. In Melbourne, 40 000 people (from a population of 200 000) turned up to a free public banquet, and the ensuing riot when the food ran short forced the prince to cancel his appearance.

Sydney was determined to outdo its rivals in the lavishness of its welcome and its display of pride in the city and its citizenry. At the welcome parade, there were no less than four elaborate triumphal archways for the prince to pass under, one nearly 20 metres high. Over the next few weeks, there were banquets, concerts, balls and fireworks.

And at three o'clock on Saturday, 22 February, there was a visit to the Australian Museum. It was a great public honour for the museum to host the prince and, along with a group of trustees, Gerard Krefft (who had only recently arrived at the museum) was able to spend 'a quiet hour' showing him around the exhibits.

The prince seems to have had an interest in Australia's natural history, but Krefft and the trustees wanted to make sure the visit was truly memorable. Snakes were Krefft's special interest, and when the official party entered the new College Street (Barnet) wing, Krefft had two cases ready. In one were live snakes, in the other two live mongooses along with a bag holding another pet mongoose belonging to Colonial Secretary Sir Henry Parkes.

The chaotic scene that followed was almost certainly memorable for all who witnessed it. In a series of fights staged to demonstrate the purported immunity of a mongoose to snakebite, the three animals were each time triumphant, with the first poor snake killed in only seconds. The second mongoose (said to be from Timor) was released to kill a number of small frogs, and then Mr Parkes' Ceylonese mongoose – which ran around the gallery and then, frightened by the crowd of people, tried to escape – went up against a large (unidentified) snake. In a dramatic confrontation, the mongoose bit the snake, which coiled itself around the little animal, attempting to suffocate it. The mongoose struck again, however, and the exhausted snake 'soon expired'. Perhaps picking his way through the dead snakes and frogs littered across the gallery floor, Prince Albert was then reported to have politely 'inspected the whale of the Museum', expressed his thanks to Mr Parkes and Mr Krefft, and returned to Government House.

Just two weeks later, on 12 March, disaster struck the prince's highly choreographed visit, when there was an assassination attempt during a picnic at Clontarf beach in Sydney. The lightly injured prince (the bullet had glanced off his ribs, inflicting only a slight wound) cut short his tour and returned to England in early April.

On leaving the colony, the prince took with him a selection of the local wildlife including live wombats, parrots, two emus, and 'a pair of large and very tame kangaroos'. He was even presented with one of the museum's mongooses that had taken part in the snake-killing demonstration. The prince and his new menagerie all headed back to England aboard the steamship HMS *Galatea*.

A NEW MANTA RAY

The coming of a new species of manta ray to Sydney Harbour shortly after the royal tour was fortuitous for Krefft. Keen to build on the prince's visit to the museum and cement a place in Alfred's memories of Australia – as well as to bolster his own scientific status and prestige in Europe's circles of power and influence – Krefft proposed naming the new ray *Manta alfredi*, or the Alfred Manta. Krefft then went even further, staging a somewhat odd but expressive set of four photographs to send to the prince with the manta ray news.

The photographs show Krefft in a variety of poses with his fish find. Deceptively casual, they are carefully composed to show both Krefft and the ray from all angles. Both top and bottom and each side of the fish are illustrated in the images – as is proper for scientific identification purposes.

There are three different views of Krefft. Ostensibly present to give the huge manta ray a human scale, he is seen in shirtsleeves looking at the camera; with jacket and top hat sitting reading a newspaper; and standing, leaning on the wall in a casual pose reading the same newspaper. This is Krefft as a coolly confident colonial scientist with the great worth of his scientific work on literal display; as comfortable in shirtsleeves (the scientific worker) as he is in formal attire (the scientific thinker and writer).

PHOTOGRAPHY REMAKES THE MUSEUM

Krefft, as we have seen, was well aware of the power of imagery to not only transparently document specimens but also artfully represent them in specific poses and settings. The carefully stage-managed images that Krefft and Barnes made for the prince are among the best examples of the crossover potential of photography at the museum, with its appeal to both aesthetic and scientific values, and its role in creating and projecting a new model and new imagery of the modern scientific man and his museum.

Photography was a useful tool to document scientific work. More than that, in the hands of a canny communicator like Krefft it could create and disseminate images of the modern scientist, demonstrate the progress of colonial science, and help us understand the role of museums within the complex mix of identity, empire, nature and science.

The Australian Museum at the end of the 19th century is a rare site where we can see the coming together of art, science and nature in the changing face and the expanding spaces, role and influence of public science in colonial Sydney. Using the museum's own archival photography collections, the story can be graphically told through the work that Gerard Krefft, Edward Ramsay and Henry Barnes did there, and through the very special collection of glass plate photography they left behind. The careful images they captured are a lively demonstration of the practice, usefulness and rapid spread of the new technology of photography, and its integral part in creating the new museology and the new museum.

Pages 181, 182–83 and opposite: After the ray was photographed for the sequence of images shown here, work on preserving the huge Alfred Manta was undertaken by taxidermist Charles Tost and was said to have required three weeks to complete. It is still held in the museum's ichthyology (fish) collection. The ray is also known as the 'Devil Ray', and a description of the specimen on display in the gallery in 1868 emphasises its menace, monstrosity and size: 'This huge brute is fourteen feet across and eight feet from the mouth to the tail – which, by-the-bye, is short and slight, like a ladies' riding whip. The monster has a mouth about two feet long, with very small teeth round the edge of the wide but by no means deep orifice. Its head is hardly perceptible' (*Sydney Mail*, 24 October 1868, p. 9).

ACKNOWLEDGMENTS

The writing and production of this book was supported by a grant from the Australian Museum Foundation. I thank them and Kim McKay, Director and CEO, and Russell Briggs, Director, Engagement, Exhibitions and Cultural Connection, for their support of this book and the *Capturing Nature* exhibition.

This book relies on many years' prior work by a large and dedicated team at the Australian Museum. In the Museum Archives, I owe a huge debt to the work of archivists Rose Docker, Patricia Egan, Elizabeth McKinnon, Sue Myatt and Prue Walker. In Collections I thank Cameron Slatyer, Sandy Ingelby, Harry Parnaby, Mark Eldridge, Mark McGrouther, Amanda Hay, Sally Reader, Stephen Keable, Derek Smith, Jodi Rowley, Glenn Shea, Stephen Mahoney, Matt McCurry and Leah Tsang for their advice and fact-checking. From Conservation I thank Colin Macgregor, Sheldon Teare and Michael Kelly. In Exhibitions I thank Gill Scott, Elliott Cole, Amanda Teer and Aaron Maestri for their commitment to this project and their creativity and hard work to bring the exhibition to life. Serafina Froio was an early supporter of the book. Reproductions and photography are by James King.

I especially want to thank Vanessa Low for help with photography selections, for coming to grips with 19th-century photography techniques to write the chapter on photographic processes, and for her enthusiastic engagement with the project.

Rose Docker researched the Wellington Caves dig and Patricia Egan took on the history of taxidermy.

This book owes a huge debt to Australian photo historians Alan Davies, Gael Newton, Jane Lydon and Judy Annear (among others). Joining this circle is Kathleen Davidson with her groundbreaking work on museum photography in colonial Australia, in particular her work on Gerard Krefft.

Historians of the Australian Museum including Ronald Strahan, Matthew Stephens, Jenny Nancarrow and numerous historians of individual collections, objects and collectors have made this history possible. For the wider history of Australian museums and natural history, I have relied on the work of Peter Hobbins, Anne Coote, Libby Robin and Jude Philp. Colleagues at other museums have been generous and helpful. From the Macleay Museum in Sydney I thank Jan Brazier; from Museums Victoria I thank Nik McGrath.

The team at NewSouth Publishing have been wonderful to work with, and I thank publisher Elspeth Menzies, project manager Emma Hutchinson, Hamish Freeman and Klarissa Pfisterer for design work, and editor Diana Hill.

AUTHOR BIOGRAPHY

VANESSA FINNEY is a curator
and archivist with a special interest
in museums and natural history in
colonial Australia. Her first book,
*Transformations: Harriet and Helena
Scott, colonial Sydney's finest natural
history artists*, was published by
NewSouth Publishing in 2018.
At the Australian Museum she heads
the country's oldest and largest
specialist natural history archives
and rare books collection.

LIST OF ILLUSTRATIONS

Note on artworks

Unless noted, all original images are held by the Australian Museum Archives and come from the Australian Museum series AMS 351, Large Format Negatives (V) collection. For reproduction requests, contact the Australian Museum Archives. Except where stated otherwise, all works are recent scans by James King of the original black-and-white glass plate negatives. Collection registration numbers were scratched directly onto the glass plates, sometimes many years after they were processed. The numbers (in the form 'Vxxxx') were scratched onto either side of the glass plate, so sometimes appear in reverse in the printed image.

Cover

front cover, V0404; back cover clockwise from top left, V0345, V0428, V0254, V1025, V0463.

Endpapers

front endpaper from left to right, V0708, V0416; back endpaper from left to right, V1714, V1732.

Front matter

pp. ii–iii, V0304; p. iv, V1714; p. v, V0151; p. vi, V0463; p. vii, V1025.

Introduction: The Museum and the Mechanical Eye

p. viii, V1292; p. 1, V0432; p. 3, V1034; pp. 4–5, V0302; p. 6, V1533.

**Part I:
Time, Place and People**

pp. 8–9, V0006.

**Chapter One:
New Visions of the
Natural World**

p. 10, V0708; p. 11, V0015; p. 13, V0961; pp. 14–15, V1510; p. 16, V0443; p. 17, V0450; p. 18 clockwise from top left, V0896, V1404, V0856, V0815, V0772, V0853; p. 19 from top to bottom, V0589, V0659, V0697; p. 20, V0783; pp. 22–23 clockwise from top left, V1061, V1063, V1097, V1442, V1602, V1087; p. 24, V0417; p. 26, illustration reproduction copyright © Cambridge University Library, Darwin Correspondence Project, 'Letter no. 8959', letter from Johann Louis Gerard Krefft to Charles Robert Darwin, 12 July 1873; p. 29, V0226; p. 30, plate III from Gerard Krefft's *Snakes of Australia*, Australian Museum Research Library; p. 31, V0352;

p. 33, V0102; p. 34, V1853; p. 37, V0706; p. 38 from top to bottom, V344, V345; p. 39, V0343; pp. 40–41 from left to right, V0431, V0340, V0342.

**Chapter Two:
The Men Behind the Images**

p. 42, V1420; p. 43, V0283; p. 44, V0629; p. 45, V2504; p. 46, copyright © Trustees of the Natural History Museum London, Albert Günther Photographic Album; p. 47, V0409; p. 48, illustration from Australian Museum Archives MA0118; pp. 50–51 from left to right, V1224, V1223, V1225; p. 52 from top to bottom, V0215, V0216; p. 54, V0285; p. 57 from top to bottom, V0254, V0255, V0256; p. 58, V0136; p. 61, V0020; p. 62, V0061; p. 63, V0060; p. 65 clockwise from top left, V1153, V1154, V0501, V0500; p. 66, V0037; p. 67, Australian Museum Archives, AMS512_01; p. 68, V0068; p. 70 top (panorama), AMS438, Papers Relating to the Exploration of Caves and Rivers, 1869–1882, and bottom (campsite), Australian Museum Archives, AMS512_19; p. 72 from top to bottom, V0155, V0154; p. 73 from top to bottom, V0262, V0151; p. 74, V0096; p. 76, V404; p. 77, Australian Museum Archives AMS160, original photographer unknown, copied and enlarged by Appleby Studio photographers, 1907; p. 79 from top to bottom, V1278, V1279, V1287.

**Chapter Three:
Making and Managing
the Collections**

p. 80, V1441; p. 81, V1921; p. 82, V1339; p. 85, V0416; p. 86 from top to bottom, V0351, V0348; pp. 88–89, V1809; p. 90 from top to bottom, V0882, V0883;

p. 91, V1834; p. 92, V0442; p. 93, V0660; p. 94, V0284; p. 95, V0323; pp. 96–97, V1006; p. 98, V11865; p. 100, Australian Museum Archives, Australian Museum Photographic Albums, AMS421_1, AMS421_2, AMS421_7; p. 102, V0835; p. 103, Australian Museum Archives, photographic negatives V register, volume 1, AMS 569/117; p. 104, V0003; p. 105, V0004; p. 106, V0458; p. 108 from top to bottom, V457, V0456; p. 109, illustration from *Illustrated Sydney News*, 15 May 1880; p. 110, Australian Museum Archives, Australian Museum Photographic Albums, AMS421_1, p. 93; p. 113, Australian Museum Photographic Albums, AMS421_10, pp. 20, 21; p. 115, V1271; p. 116, Australian Museum Archives, Henry Barnes Photographic Album, AMS512; p. 117 top row from left to right, middle row left, Australian Museum Archives, Henry Barnes Photographic Album, AMS512, and middle row right, bottom row left to right, Australian Museum Photographic Albums, AMS421_21.

Part II:
Artisans and Technicians

pp. 118–19, V0011.

Chapter Four:
The Australian Museum
Photographer, 1857–1893

p. 120, V1718; p. 121, V1732; p. 123, Australian Museum Archives, Curators' Reports to the Trustees, AMS24/1881; p. 124 clockwise from top left, V1928, V1931, V1933, V1932; p. 126, illustration from *Marion's Practical Guide to Photography*, Marion and Co., London, 1885, p. 249; p. 127 from left to right, V1171, V1173, V1187; p. 128 top, V1297, and bottom from left to right, V1300, V1301; p. 130, illustration from the History Collection, Alamy Stock Photo, image ID: J2YARH; p. 131, V0032; p. 132, V922; p. 133, V1936; pp. 134–35 from left to right, V0336, V0337; p. 136, V1530; p. 137, V1554; p. 138–39 from left to right, V1696, V1459, V1471; p. 140, V0007; p. 141, illustrations from John Thomson (ed.), *A History and Handbook of Photography: translated from the French of Gaston Tissandier*, Sampson Low, Marston, Searle & Rivington, London, 1878; p. 142, V0403; p. 143, V1983; p. 144, V1417; p. 145, V0920; p. 146 top and bottom, V0625; p. 147, illustration from *Buch der Erfindungen, Gewerbe und Industrien*, Otto Spamer Publishing House, Leipzig and Berlin, 1864–1867, Alamy Stock Photo, image ID: FF6W3R; p. 149 from top to bottom, V0199, V0185.

Chapter Five:
The Art of Taxidermy
and Articulation

p. 150, V1401; p. 151, V2014; pp. 152–53, V0789; p. 155, V0788; p. 156, V0705; p. 157, illustration from *Glenn Innes Examiner and General Advertiser*, 2 June 1885, accessed via Trove <trove. nla.gov.au/newspaper/page/23888962> (accessed 12 November 2018); p. 159 from top to bottom, V0428, V0429; p. 160, V1922; p. 163, V1470; pp. 164–65, V1429; p. 166, V1428; p. 168, V1419; p. 171, V1427; p. 172, V1260; p. 175 from top to bottom, V0300, V0329; pp. 176–77 from left to right, V0332, V0331, V0330; p. 178, V1496; p. 181, V0017; pp. 182–83, V0016; p. 184 from top to bottom, V0018, V0019.

NOTES

Introduction:
The Museum and
the Mechanical Eye

1   'The Improvements Effected in
    Modern Museums in Europe and
    Australia', a paper read before the
    Royal Society of New South Wales
    on 5 August 1868, *Journal and
    Proceedings of the Royal Society of
    New South Wales*, vol. 2, 1869, p. 24.

Chapter One:
New Visions of
the Natural World

1   Anonymous author, *The Sydney
    Gazette and New South Wales
    Advertiser*, 29 June 1827, p. 2.

2   George Bennett, *Wanderings in
    New South Wales*, vol. 1, Richard
    Bentley, London, 1834, pp. 67–69.
    Bennett described the collection
    in detail and declared: 'the
    commencement of the public
    museum is excellent'. Bennett
    was appointed curator of the
    Australian Museum in 1835.

3   *A Catalogue of the Specimens of
    Natural History and Miscellaneous
    Curiosities Deposited in the
    Australian Museum*, printed by
    James Tegg and Co., Sydney, 1837.

4   *Empire*, 16 May 1868, p. 5. From a
    column written by (anonymous
    author) 'The Flaneur in Sydney'.

5   Letter from Gerard Krefft to Henry
    Parkes, 23 September 1873. Sir
    Henry Parkes Correspondence,
    vol. 20, State Library of New South
    Wales, ML A890. In Ronald Strahan,
    *Rare and Curious Specimens: an
    illustrated history of the Australian
    Museum 1827–1979*, Australian
    Museum, Sydney, 1979, p. 35.

6   AW Scott, *Australian Lepidoptera
    and their Transformations: drawn
    from the life by Harriet and Helena

Scott; with descriptions, general
and systematic, by AW Scott*, vol. 1,
John van Voorst, London, 1864; and
vol. 2, John van Voorst and the
Australian Museum, London and
Sydney, 1898.

7   William Henry Fox Talbot, *The
    Pencil of Nature*, Longman, Brown,
    Green and Longmans, London,
    1844–46, plate viii, 'A Scene in a
    Library'.

8   See John Hannavy, 'Roger Fenton
    and the British Museum', *History of
    Photography*, vol. 12, no. 3, 1988,
    pp. 193–204.

9   Other British museums were also
    making use of photography. These
    included the South Kensington
    Museum (now known as the
    Victoria and Albert Museum) in
    London. By 1880 this museum had
    amassed 50 000 photographs of
    representative works of Western
    and non-Western art and culture,
    available to researchers and artists.
    See Phillip Prodger, *Victorian
    Giants: the birth of art
    photography*, National Portrait
    Gallery, London, 2018, p. 170.
    The Museum of Economic Botany
    at Kew Gardens was also collecting
    and displaying photography. See
    Caroline Cornish, 'Collecting
    Photographs, Constructing
    Disciplines: the rationality and
    rhetoric of photography at the
    Museum of Economic Botany', in
    Elizabeth Edwards and Christopher
    Morton (eds), *Photographs,
    Museums, Collections*: between art
    and information, Bloomsbury
    Academic, London, 2015,
    pp. 119–38. Both these examples,
    however, demonstrate photographs
    being externally collected, not
    made, by institutions.

10 Kathleen Davidson, *Photography,
    Natural History and the
    Nineteenth-Century Museum:
    exchanging views of empire*,
    Routledge, Oxford, 2017. Davidson's
    wonderful book is the seminal
    account of this moment in the
    colonial context.

11 Preface to Gerard Krefft, *The
    Snakes of Australia: an illustrated
    and descriptive catalogue of all the
    known species*, Thomas Richards,
    Government Printer, Sydney, 1869.

12 Trust Minutes, September 1878.
    Australian Museum Archives,
    AMS01.

13 Trust Minutes, October 1869.
    Australian Museum Archives,
    AMS01.

14 Caroline Cornish states that
    photographs were 'widely used
    by colonial commissioners in
    international exhibitions to
    illustrate the resources of their
    respective colonies and to present
    themselves as modern and
    progressive, all in an attempt to
    attract emigrants and investment',
    in Cornish, 'Collecting Photographs,
    Constructing Disciplines', p. 123.
    At the London 1862 International
    Exhibition, 600 photographs from
    all Australian states were displayed.

15 *Official Record of the Melbourne
    International Exhibition, 1880–1881*,
    Mason, Firth & McCutcheon,
    Melbourne, 1882, p. cxi.

16 Curators' Reports. Australian
    Museum Archives, AMS24/8/1881.

17 Curators' Reports. Australian
    Museum Archives, AMS24/6/1895.

18 Gerard Krefft, 'Australian
    Crocodiles', *The Australasian*,
    vol. 3, no. 69, 27 July 1867, p. 8.

19 These photographs are not
    held with Krefft's related

correspondence at the Natural History Museum, London. It appears that photographs were routinely removed from correspondence and filed elsewhere (as they were at the Australian Museum).

20 Letter from John Blaxland (donor) to Gerard Krefft, 13 January 1873. Australian Museum Archives, Letters Received, AMS07, C.30.73. Johnstone gave the crocodile to Blaxland, who in turn gave it to the museum. Krefft spelt Johnstone's name incorrectly in his first description. The spellings *crocodylus*, *crocodilus*, *johnsoni*, *johnsonii*, *johnstoni* and *johnstonii* are all now acceptable, since each version of the name has been in general usage.

21 Glenn M Shea, Cecilie A Beatson and Ross Sadlier, 'The Nomenclature and Type Material of *Crocodylus johnstoni* (Krefft, 1873)', *Records of the Australian Museum*, vol. 68, no. 3, 2016, pp. 81–98. Australian Museum Archives.

**Chapter Two:
The Men Behind
the Images**

1   William Blandowski, *Australien in 142 photographischen Abbildungen ...*, was published in Germany in 1862. For an English translation see Harry Allen (ed.), *Australia: William Blandowski's illustrated encyclopaedia of Aboriginal Australia*, Aboriginal Studies Press, Canberra, 2010 ('Portrait of Yarree-Yarree Aborigines based on a photograph by Blandowski, 1857' appears on p. 45).

2   John Kean has written a fascinating account of the expedition, and of Gerard Krefft's growing scientific confidence through collecting, trading and observation. J Kean, 'Observing Mondellimin: or when Gerard Krefft "saved once more the honour of the exploring expedition"', *Proceedings of the Royal Society of Victoria*, vol. 121, no. 1, 2009, pp. 109–28.

3   Letter from Gerard Krefft to Frederick McCoy, 28 January 1858. Outward Letterbook 2/1. Museum Victoria Archives.

4   Kean, 'Observing Modellimin', p. 110.

5   Gerard Krefft's Blandowski expedition illustrations are held at the Natural History Museum, Berlin.

6   See Paul Humphries, 'Blandowski Misses Out: ichthyological etiquette in 19th-century Australia', *Endeavour*, vol. 27, no. 4, 2003, pp. 160–65, for a full account of Blandowski's travails in getting his scientific work recognised in Melbourne when his attempts to 'honour' members of the Philosophical Institute of Victoria backfired spectacularly.

7   John Kean states that the men's 'inability to manage the relationship for mutual benefit saw the material they collected dispersed and the impact of their achievement evaporate', in Kean, 'Observing Modellimin', p. 109.

8   For full coverage of early Australian palaeontology and the importance of the Wellington Caves deposits in early debate, see Peter Minard, '"Making the 'Marsupial Lion'": bunyips, networked colonial knowledge production between 1830–59 and the description of *Thylacoleo carnifex*', *Historical Records of Australian Science*, vol. 29, no. 2, 2018, pp. 91–102.

9   Letter from Charles Darwin to Gerard Krefft, 17 July 1872. State Library of New South Wales, MLMSS 5828. In closing, Darwin writes, 'if you will allow me to say so, I have long respected your able & indefatigable labours in the cause of natural science'.

10 Letter from Gerald Krefft to his brother William in Braunschweig (Brunswick), Germany. Undated fragment, but probably 5 November 1872. Correspondence held by the Krefft family, Hamburg, Germany; copied by Jenny Nancarrow and translated by Bodo Matzik in 2000.

11 *Visitors' Guide to Sydney*, William Maddock, Sydney, 1872.

12 *Australian Town and Country Journal*, 23 December, 1871, p. 13.

13 Jenny Nancarrow, 'Gerard Krefft: a singular man', *Proceedings of the Royal Society of Victoria,* vol. 121, no. 1, 2009, p. 150.

14 Strahan, *Rare and Curious*, gives an exhaustive account of the enquiry and its findings, pp. 33–35.

15 *Evening News*, 22 February 1881.

16 In 1889, the discovery was at the top of a list published in the leading British scientific journal *Nature*, describing the most important zoological discoveries of the past twenty years. Quoted in Matthew Stephens, 'The Australian Museum Library: its formation, function and scientific contribution, 1836–1917', PhD thesis, University of New South Wales, 2013, p. 210. See also 'Twenty Years', *Nature*, vol. 1, no. 16, 7 November 1889, p. 3.

17 Libby Robin, *How a Continent Created a Nation*, UNSW Press,

Sydney, 2007, p. 39. Perhaps Gerard Krefft's sense of urgency also stemmed from a more local colonial competitiveness: the fish had first been pointed out to him by William Forster some years earlier, and Krefft had dismissed his observations.

18 Australian Museum trustee and geologist William Branwhite Clarke later claimed that William Forster had intended to send the specimen to him, not to Gerard Krefft. Stephens, 'The Australian Museum Library', pp. 213–14.

19 Since the fish did not fit any known categories (indeed it appeared to cross or collapse firm category boundaries) and was often categorised as 'bizarre' or 'monstrous' in contemporary accounts.

20 Letter from PL Sclater, secretary of the Zoological Society, quoted by Gerard Krefft. *Sydney Morning Herald*, 9 June 1870, p. 3.

21 The photographs suggest that the local Aboriginal women were active collaborators in the museum's early *Ceratodus* hunts. A decade later, Indigenous people from the same area assisted British scientist WH Caldwell with his platypus observations, which led to the 'discovery' that platypuses lay eggs. See Libby Robin, 'Paradox on the Queensland Frontier: platypus, lungfish and other vagaries of nineteenth-century science', *Australian Humanities Review*, no. 19, 2000.

22 *Sydney Morning Herald*, 20 December 1871.

23 Letter to the museum trustees from Henry and Robert Barnes,

and Charles and Jane Tost, 3 September 1868. Australian Museum Archives, AMS07, P:10:68.

24 Australian Museum Annual Report, 1897, p. 2. Australian Museum Archives.

25 'Mr Krefft's Report on the Fossil Remains Found in the Caves of Wellington Valley', *Sydney Morning Herald*, 18 December 1866, p. 5.

26 'Mr Krefft's Report'.

27 'Mr Krefft's Report'.

28 'Mr Krefft's Report'.

29 Report of Gerard Krefft to the trustees of the Australian Museum, 7 October 1869, in 'Exploration of the Caves and Rivers of New South Wales', presented to Parliament by command, T Richards, Government Printer, Sydney, 1882.

30 Letter from Richard Owen to Gerard Krefft, 26 February 1870, f. 30–32 and 22 April 1870: 27, f. 39–41. Owen correspondence, Natural History Museum, London, quoted by Davidson in *Photography, Natural History*, p. 94.

31 Manuscript prepared by Gerard Krefft, 1874. A copy is held in Australian Museum Archives, AMS139/4/78. The location of the original is unknown.

32 Charles Darwin, *On the Origin of Species*, 6th edn, John Murray, London, 1872, pp. 236–37.

33 *Records of the Australian Museum*, vol. 11, no. 9, May 1917. Australian Museum Archives.

Chapter Three: Making and Managing the Collections

1 'The Improvements Effected in Modern Museums in Europe and Australia', a paper read before the Royal Society of New South Wales on 5 August 1868, *Journal and Proceedings of the Royal Society of New South Wales*, vol. 2, 1869, p. 15.

2 *The Empire*, 13 June 1861, p. 4.

3 *Australian Dictionary of Biography* <adb.anu.edu.au/biography/masters-george-4166> (accessed 6 July 2018).

4 Letter from Gerard Krefft to JE Gray, 26 March 1869. British Museum (Natural History) Archives.

5 Australian Museum Annual Report, 1857, p. 1. Australian Museum Archives.

6 British historian Elizabeth Edwards has written extensively about anthropology, photography and museums. See, for example 'Material Beings: objecthood and ethnographic photographs', *Visual Studies*, vol. 17, no. 1, 2002, pp. 66–75. In the Australian context, see the work of historian Jane Lydon.

7 Curator Edward Ramsay reported that it was 8 feet long and 11 feet from the tip of the dorsal fin to the tip of the ventral fin.

8 'News of the Day', *Sydney Morning Herald*, 13 December 1882, p. 9.

9 Gilbert P Whitley, 'Studies in Ichthyology. No. 4', *Records of the Australian Museum*, vol. 18, no. 3, 1931, p. 131. Australian Museum Archives. This article contains a description, attributed to Edward Ramsay, of all the 1882 and 1883 sunfish, see pp. 131–32.

10 *Sydney Morning Herald*, 16 December 1871, p. 7.

11 *Sydney Morning Herald*, 20 December 1871.

12 Marcus Strom, 'Natural History Museum of London finds *Sydney Morning Herald* from Australia Day

1883 inside sunfish', *Sydney Morning Herald*, 23 September 2016.

13   The fate of the sunfish taken at Botany Bay is unknown.

14   Transcript of the Proceedings of the New South Wales Legislative Assembly, *Sydney Morning Herald*, 11 March 1859, p. 5.

15   This new rigid denomination of species by 'type' ties in with the falling status of composite scientific illustrations and the rise of one-off photographic images of particular specimens.

16   Peter Hobbins, *Venomous Encounters*, Manchester University Press, Manchester, 2017, p. 22. The colony's imported sheep and cattle are a notable exception – sheep-stealing was punishable by death.

17   The Blandowski expedition collected 750 mammal specimens and 29 mammal species.

18   For a history of the colonial/animal matrix, see Hobbins, *Venomous Encounters*. Hobbins takes a fascinating look at the history of snakebite, snake vivisection and venom studies in Australia.

19   Gerard Krefft, *The Mammals of Australia*, Thomas Richards, Government Printer, Sydney, 1871, plate xiv, 'Spiny Ant-eater'.

20   Trust Minutes, 6 March 1879. Australia Museum Archives, AMS01.

21   See Edwards and Morton, *Photographs, Museums, Collections* for wider, comparative discussion of museum photography collections.

Chapter Four:
The Australian Museum Photographer, 1857–1893

1   Letter from Nikolai Miklouho-Maclay to Edward Ramsay, 10 September 1878, Australian Museum Archives, AMS07; H40:78. Russian naturalist and anthropologist Miklouho-Maclay had just arrived in Australia and was given permission to use the 'Photographique atelier' of the museum to photograph objects 'for the purpose of illustrations in connection with my research which I shall publish in Europe, as well as in Sydney'.

2   Wilson Bentley's book, *Snow Crystals*, McGraw-Hill, New York, 1931, featured 2500 crystal photographs.

3   The image appeared in Robert Etheridge, jnr, 'Additions to the Middle Devonian and Carboniferous Corals in the Australian Museum', *Records of the Australian Museum*, vol. 4, no. 1, 1902. Australian Museum Archives.

4   Con Tanre, *The Mechanical Eye*, Macleay Museum, University of Sydney, Sydney, 1977, p. 38. None of George Goodman's prints seem to have survived.

5   Tanre, *The Mechanical Eye*, pp. 107–23.

6   Although the first Amateur Photographic Society in Australia was formed in Sydney in 1872, by 1886 it still had only 140 members in the whole of New South Wales.

7   William Saville-Kent, 'Notes on New and Little Known Australian Madroporaceae', *Records of the Australian Museum*, vol. 1, no. 6, 1891, pp. 123–24. Australian Museum Archives.

8   The Australian Museum transitioned from wet to dry plate photography in 1881. Notes from the Curator's Report on 5 July 1881 list the desired purchase of almost 300 dry plates from Mawson & Swan, noting that half would be '15 times as quick as wet plates' and the other half '20 times'. Curators' Reports. Australian Museum Archives, AMS24/8/1881.

9   Robert Hunt, *A Manual of Photography*, Richard Griffin and Company, London, 1857, p. 108.

10   *Evening Journal*, 23 June 1870, p. 2.

11   Richard Owen, 'On the Fossil Mammals of Australia. Part 1. Description of a mutilated skull of the large marsupial carnivore (*Thylacoleo carnifex*, Owen), from a calcareous conglomerate stratum, eighty miles S.W. of Melbourne, Victoria', *Philosophical Transactions of the Royal Society*, 1859, vol. 149, pp. 309–22.

12   WS Macleay, 'The Native "Lion" of Australia', letter to the Editor, *Sydney Morning Herald*, 1 January 1859, p. 5.

13   G Krefft, 'On the Dentition of *Thylacoleo carnifex* (Owen)', *Annals and Magazine of Natural History*, vol. 18, 1866, pp. 148–49.

Chapter Five:
The Art of Taxidermy and Articulation

1   Anne Coote, 'Science, Fashion, Knowledge and Imagination: shopfront natural history in 19th-century Sydney', *Sydney Journal*, vol. 4, no. 1, 2013, pp. 1–18. See p. 6 of this article for a list of women who worked alone or as part of a family taxidermy business. In 1996 the Macleay Museum at Sydney University held an exhibition celebrating women taxidermists: *Most Curious and Peculiar: women taxidermists in colonial Sydney*.

2   Rose Docker, 'The Queerest Shop in Australia', *Explore* [magazine of the Australian Museum], vol. 34, no. 2, 2012.

3   Anne Coote has described Sydney's commercial natural history trade in detail in Coote, 'Science, Fashion, Knowledge and Imagination'.

4   Letter from Gerard Krefft to Richard Owen, British Museum, 21 October 1864 (letter 191). Outward Letter Books. Australian Museum Archives, AMS06.

5   Frederick McCoy jubilantly announced the acquisition in the *Melbourne Argus*, 20 June 1865. For a full account of the purchase and the Melbourne reaction see Joan M Dixon, 'Melbourne 1865: gorillas at the museum', <museumsvictoria.com.au/history/gorillas.html> (accessed 2 September 2018).

6   Description of the Osteology Gallery, *Sydney Mail*, 10 October 1868.

7   Letter from Sutherland Sinclair to E Gerrard, London, 10 May 1883 (letter 121). Outward Letter Books, Australian Museum Archives, AMS06.

8   'Museum Brings Nature to Life', *Sydney Morning Herald*, 14 September 1887, p. 5.

9   Victorian debates around gorillas were always heavily racialised.

10 'Museum Brings Nature to Life'.

11 W.H.Y., 'Evolution: An Address to the Gorilla in Sydney Museum', *Sydney Mail and New South Wales Advertiser*, 9 December 1903, p. 1502.

12 Australian Museum Annual Report, 1869, p. 1. Australian Museum Archives.

13 Figures taken from Australian Museum Annual Reports for the decade 1870–90. Australian Museum Archives.

14 Edward Ramsay, *Notes for Collectors: containing hints for the preservation of specimens of natural history*, Australian Museum Miscellaneous Publications, no. 3, 1887. Gerard Krefft, *Catalogue of Mammalia in the Collection of the Australian Museum*, Sydney, 1860.

15 Ramsay, *Notes for Collectors*, p. 8.

16 Ramsay, *Notes for Collectors*, p. 17.

17 Gerard Krefft, *Catalogue of Mammalia in the Collection of the Australian Museum*, 2nd edn, Sydney, 1864, p. 136.

18 William Swainson, *Taxidermy*, Longman, London, 1840, p. 29.

19 Letter to OE Janson Esq, 11 December 1894. General Order Book. Australian Museum Archives, AMS065, p. 180.

20 Jude Philp, 'Stuffed', in J Philp, T Gill, R Blackburn and L Lui-Chivizhe (eds), *Stuffed, Stitched and Studied: taxidermy in the 19th century*, Macleay Museum, University of Sydney, Sydney, 2015 p. 23.

21 Arsenic soap still leave specimens vulnerable to attacks from bacteria and fungi, since both of these are resistant to arsenic.

22 See Fernando Marte, Amandine Pequignot and David W Von Endt, 'Arsenic in Taxidermy Collections: history, detection, and management', *Collection Forum*, vol. 21, nos 1–2, 2006, pp. 143–50.

23 At the time of writing the 'Merchants and Museums' Australian Research Council project is running at the Australian Museum, Wollongong University, Melbourne Museum and the University of Sydney with the aim of understanding the global scale of the commercial trade in natural history specimens across the 19th century. 'Reconstructing Museum Specimen Data through the Pathways of Global Commerce', project ID: LP160101761.

24 For a full history of the habitat group in museums see Karen Wonders, 'Exhibiting Fauna – from spectacle to habitat group', *Curator*, vol. 32, no. 2, June 1989, pp. 131–56.

25 Australian Museum Annual Report, 1871. Australian Museum Archives.

26 Today, the museum's taxidermists enlist Dermestid beetles, *Dermestes maculatus*, for the task.

27 Jude Philp, 'Taxidermy', in J Philp, T Gill, R Blackburn and L Lui-Chivizhe (eds), *Stuffed, Stitched and Studied: taxidermy in the 19th century*, Macleay Museum, University of Sydney, Sydney, 2015, p. 17.

28 Thomas Brown, *The Taxidermist's Manual, or, the Art of Collecting, Preparing, and Preserving Objects of Natural History: designed for the use of travellers, conservators of museums and private collectors*, A Fullarton, London, 1853, p. 95.

29 Trust Minutes, 5 October, 1876. Australian Museum Archives, AMS01.

30 The photograph Krefft sent to Gray appears on p. 175 (top, *Mesoplodon densirostris*). Gray's description 'On the Skeleton of *Dioplodon sechellensis* in the Australian Museum at Sydney' appeared in the *Annals and Magazine of Natural History*, vol. vi, Taylor and Francis, London, 1879, pp. 343–44.

31 Letter from Gerard Krefft to John Gray, 1871, Australian Museum Archives, AMS07.

INDEX

Bold page numbers refer to illustrations.

3D scanning 109

Aboriginal Australians 49, 53, 63, 173
Act for the Protection of Animals 1881 109
'Address to the Gorilla in Sydney Museum' 157
Adelaide, SA 180
Ailuroedus crassirostris **79**, 162, **163**
Albert, Prince Alfred Earnest 180, 185
Alectis ciliaris **18**, 19
Alfred Manta 180, **181–84**, 185
Alligator Gar 21, **23**
amphibians 60, 63
Amphichaetodon howensis 132, **133**
analytical spectroscopy 167
Angas, George French 15, 87
anglerfish **19**
animal ethics 107, 109
Animal Locomotion (1887) 27
Animal Research Act 1985 109
ankerite **91**
Antennarius striatus **19**
Arctides antipodarum **138**, 139
Arothron hispidus **18**, 19
arsenic in taxidermy 166–67
art 26–27, **92**, 112–13, 164
articulation *see also* Osteological Gallery; staging displays
    Henry Barnes' work on 6, 64, 66, 153, 179
    learning 153
    process 174, 179
    workshop 122
Aspidiotes melanocephalus 30, **30–31**
Audubon, John James 48
Australian Freshwater Crocodile **38–41**, 39–40
Australian Hall 162, 170
Australian Lungfish 60, **61–63**, 63
Australian Museum
    College Street wing 16, **16–17**, 56
    ethnographic collection 78, 92–93, **92–93**
    exterior **144**
    Fish Gallery **37**
    founding 11–15
    Long Gallery 13, **13–15**, 15, **26**, 78, 174

    Osteological Gallery v, 101, 173, 177
    'Wild Planet' Gallery 174
Australian Museum Act 1853 13
Australian Museum Report of 1869 71
Australian Pelican **88–89**
Australian Pineapplefish **18**, 19
Australien (1862) 49, 53

Banded Seaperch **18**, 19
Banded Wobbegong **10**, 13
bandicoots v, **vi**, 64, 107
Banks, Joseph 26
Barnes, Henry **45**
    as articulator 6, 64, 66, 153, 179
    as taxidermist 6, 64, 66–67, 158
    casting and displays 64, 66, 153
    commercial photography 59, 67
    death 64
    Gerard Krefft and 59
    in Wellington Caves 33, 64, 66, 71
    photography for leisure 116
    photography for the museum 6–7, 33, 45, 64, 71, 129
    retirement 64, 67
Barnes, Henry, jnr **45**, 66
Barnes, Robert **ii–iii**, v, **45**, 59, 66
batfish **20–21**
beaked whales **4–5**, 174, **175**, 179
Beatson, Cecilie 40
Bécoeur, Jean-Baptiste 167
Bennett, George 15, 59
Bentley, Wilson 125
Bettong **58**
birds **24**, 78, 107, **160**, 161, 169
    see also bowerbirds; Brown Flycatcher; cassowaries; cockatoos; Emu; Green Catbird; Greater Flamingo; Jacky Winter; Kagu; Noisy Friarbird nest and eggs; pelicans; robins; Striated Fieldwren; Superb Lyrebird
Birds of America (1827–38) 48
Black-headed Python 30, **30–31**
Blackwood, William 126
Blainville's Beaked Whale 174, **175**
Blandowski expedition
    animals collected on 49, 53, 107
    drawings 49, 53

    photography 6, 28, 32, 49
    relationships during 48–49, 53
Bluethroat Wrasse 21, **23**
blue-tongued lizards **46**
Board of Trustees 15
bones *see* articulation; Osteological Gallery
Botany Bay, NSW 99
bowerbirds **121**, 122, 162, **163**
Bridge Street, Sydney 126
British Museum
    as a model 21, 83
    donations from 154
    examining specimens 39, 66, 148, 179
    exchanges with 64
    International Fisheries Exhibition 99
    photography in 28
Broadbent, Kendall v, 92
Broad-tailed Grassbird **160**, 161
Broken Bay, NSW 139
Brown-eared Pheasant **24**
Brown Flycatcher **34**, 35
brown snakes **86**, 87
Brunswick, Germany 46
Buchan, Alexander 26
bullseye **131**
Bump-head Sunfish **96–98**, 97, 99
Burdekin River, Qld 66
Burketown, Qld 39
Burnett River, Qld 84
butterflyfish 132, **133**

Calamanthus fuliginosus albiloris **151**, 153
calotypes 130
cameras *see* photography
Cape York, Qld 158
Carcharodon carcharias 7, **8–9**, **104–105**
cassowaries **76**, 77
casting 64, 66, 153, 166
Casuarius casuarius **76**, 77
Catalogue of Echinodermata in the Australian Museum (1885) 2
Catalogue of Mammalia in the Collection of the Australian Museum (1864) 81, 151
catbirds **79**, 162, **163**

Chadwick, Robert 97
Chaeropus ecaudatus 64, 107
Cheilopogon pinnatibarbatus 21, **22**
Chelmonops truncatus **18**, 19
Chrysophrys auratus **11**, 13
cicadas 142, **143**
Clark, GE 114
Clarke, William Branwhite 15, 59
classification
    importance of 26–27
    of photographs 36, 111, 114
    of specimens 39–40, 101, 153
Cleidopus gloriamaris **18**, 19
Clontarf Beach, NSW 180
cockatoos **116**
Collection Registers 104
College Street wing 16, **16–17**, 56
Common Dolphin 174, **175**
Common Eagle Ray **44**, 45
Common Water Monitor **172**, 173
Common Wombat 56, **57**, **82**, 83
Coogee, NSW v
Cook, James 26
coral **110**, **124**, 125, **128**, 129
    see also Gorgonian Coral
Corucia zebrata **86**, 87
Cox, James 58, 67, 99
crabs **iv**, v, **100**
crocodiles **38–41**, 39–40
Crossoptilon mantchuricum **24**
crustaceans **iv**, v, **100**, **138–39**
crystals **91**
cuckoos 78
Cultural Connection 78, 92–93
Cuvier, Georges 148
Cyclochila australasiae 142, **143**

daguerreotypes 130
Daintree, Richard 28
Darling Harbour, NSW 97
Darling River, NSW *see* Blandowski
    expedition
Darwin, Charles 25–26, 55, 69, 157
    see also On the Origin of Species
Dasyrus sp. **v**
Dasyurus viverrinus 162, **164–65**
Delphinus delphis 174, **175**
Dendrolagus dorianus **6**
Dendrolagus lumholtzi 114, **115**
Denison, William 16, 53
Dense-beaked Whale 174, **175**
Dermochelys coriacea **106**, 107,
    108
'Description of a Gigantic Amphibian'
    63
Devil Ray 180, **181–84**, 185
Diamond Python **46**

Dicerorhinus sumatrensis **102**
Diprotodon **33**, 64, **65**, **68**, 87, 173
displays *see* staging displays
DNA sampling 109
dogs **109**
dolphins **3**, 94, **95**, 174, **175**
donation 83–84, 87, 99, 154, 179
Doria's Tree-kangaroo **6**
Double Drummer Cicada 142, **143**
Dromaius novaehollandiae **48**,
    50–51, 136, **137**
dry plate process 28, 130, 141
du Chaillu, Paul 154
Duke of York Islands, PNG 93

Eastern Blue-tongued Lizard **46**
Eastern Brown Snake **86**, 87
Eastern Fiddler Ray v, **vii**
Eastern Quoll 162, **164–65**
Eastern Rock Lobster **138–39**
Eastern Talma **18**, 19
Eastern Water Dragon **46**
echidnas 109, 158, **159**, 162
'Educational series of Natural History
    specimens' 64
Elizabeth Bay, NSW 174
Ellery, RLJ 27
Elops hawaiensis 21, **22–23**
Emu **48**, **50–51**, 136, **137**
Endeavour (ship) 26
Etheridge, Robert, jnr 36, 66, 78, 125
ethics 107, 109, 173
ethnographic collection 78, 92–93,
    92–93
Ethnography Hall 92
Eubalichthys mosaicus 21, **22–23**
Evening Journal 147
Evening News 59
evolution 60, 63, 78, 154 *see also*
    Darwin, Charles
exchanges
    Diprotodon cast exchange 64
    first 83, 87
    in 1885 91
    networking for 78
    of photographs 33, 35
'Exploration of the Caves and Rivers
    of New South Wales' (1882) 75
Extatosoma tiaratum **viii**, 2

Fenton, Roger 28
field collecting 84, 87, 107, 109, 161–62
    see also Blandowski expedition;
    Wellington Caves
field photography 28 *see also*
    Blandowski expedition;
    Wellington Caves

fieldwrens **151**, 153
fish **18–19**, **22–23**, 35, 60, 63, **100**, 112
    see also names of specific fish
    Fish Gallery **37**
flamingos **178**, 179
flounder **18**, 19
Flower, William Henry 148
flycatchers **34**, 35
flyingfish 21, **22**
Forster, William 60
Fossey, Dian 157
freshwater crocodiles **38–41**,
    39–40
freshwater prawns **139**
friarbird nest and eggs **142**

Gabon, Central Africa 154–55
Galeocerdo rayneri **140**
gars 21, **23**
Garden Palace, Sydney 78, 92–93
Germany 46–47, 53
Gerrard, Edward 154, 157
Giant Bandicoot v, **vi**
giant herrings 21, **22–23**
Giant Prickly Stick Insect **viii**, 2
glass plates
    exposure of **127**
    flaws in **144**, 170
    information revealed by 122
    masking of **140**, **146**, 147
    technique for using 127, 130, 140–41,
        141, 144, 147
    'V negative' collection 111
Glaucosoma scapulare **19**
Glen Innes Examiner 157
Goldie, Andrew 92
Goodman, George 126
Gorgonian Coral **120**, 122
gorillas 154, **155–57**, 157
Gould, John 48
Grant, Robert 114
grassbirds **160**, 161
Gray, John 39, 179
Greater Flamingo **178**, 179
Great Exhibition 169
Great White Sharks 7, **8–9**, **104–105**
Green Catbirds 162, **163**
Green Grocer Cicadas 142, **143**
Gubbi Gubbi people 63
Guidebook (Australian Museum) 169,
    170
Günther, Albert 63, 66

Hawaiian Giant Herring 21, **22–23**
Herbert River, Qld 39
herrings 21, **22–23**
Hill, Edward Smith 59

Hints for the Preservation of Specimens of Natural History (1887) 161
Holmes, William 87
holotypes v, 39, 103, 135, 153
human remains 173
Humpback Whale **ii–iii**, v, **176–77**
Hunt, Robert 142
Hunterian Museum 148
Huxley, Thomas 154
Hypoplectrodes nigroruber **18**, 19

Illustrated Sydney News **109**
Indigenous Australians 49, 53, 63, 173
insects **viii**, 2, 142, **143**
Intercolonial Exhibitions 64
International Exhibitions 35, 66, 92, 112, 169
International Fisheries Exhibition 99

Jacky Winter **34**, 35
Johnson, G Randall 77
Johnstone, Robert 39
Junceella sp. **120**, 122

Kagu **79**, 83
Kajikia audax 116, **118–19**
kangaroos **150**, 153 *see also* Rat-kangaroo; tree-kangaroos
Kensington, NSW 21
Kerry, Charles 170
killer whales **43**, 45
koalas 170
Kodak 127
Kogia breviceps **54**, **94**, **134–35**
Krefft, Annie 56
Krefft, Gerard
    Alfred Manta and 185
    appointed curator 6, 15, 53
    articulation and 179
    British Museum visit 28
    Catalogue of Mammalia in the Collection of the Australian Museum (1864) 79
    Charles Darwin and 26
    dismissal as curator 6, 59, 77, 99
    early life 46, 48
    Edward Ramsay and 77
    expedition with Blandowski 48–49, 53
    field collection in Murrurundi 87
    improvements to field collecting 84
    improvements to the museum 56
    instructions for taxidermy 161
    in Wellington Caves 55, 64, 69, **70**, 71, 75

naming and description of animals 39–40, 46, 63, 84
    on natural history 25
    on photography 1
    on the Marsupial Lion 148
    on the Southern Cassowary 77
    photographic program v, 6–7, 32–33, 45
    photography for leisure 116
    purchases gorilla specimens 154
    Scott sisters and 27
    Snakes of Australia (1869) 30
    vivisection by 109
    with a Leatherback Turtle **106**, 107
    with a manta ray 180, **181–84**
    with snakes and lizards **46**
    with whales **4–5**, **15**
    work on classification 101, 103
    work on reptiles 83–84
Krefft, Hermann 56
Krefft, Rudi 56

Lake Colongulac, Vic. 148
Lasiorhinus latifrons 28, **29**
Leatherback Turtle **106**, 107, **108**
leatherjacket 21, **22–23**
Leichhardt, Ludwig 66
Lettuce Coral **128**, 129
lighting for photography 136
Linnean Society 104
lithography 30
Little Bay, NSW v, 177
Little Manly, NSW 99
lizards **46**, **86**, 87
Lobotes surinamensis **19**
London 35, 148, 154, 169
    see also British Museum; Natural History Museum; Zoological Society of London
Long Gallery
    adding a third storey to 78
    in 1860 **14–15**
    in 1887 **13**
    opening of 13, 15
    passage behind cases **26**
    whale skeletons in **14–15**, 174
Long Reef, NSW 122
longtoms (fish) 21, **23**
Lord Howe Butterflyfish 132, **133**
Lucas, Augustin 126
Lumholtz's Tree-kangaroo 114, **115**
lungfish 60, **61–63**, 63
lyrebirds **42**, 45, **168**, 169

Macarthur, William 15
McCoy, Frederick 26, 49, 154

Macleay, Alexander 12
Macleay, George 15, 58
Macleay, William John 84
Macleay, William Sharp 15, 26, 58, 77, 148
Macleay Museum 84
Macrobrachium sp. **139**
Macropus rufus **150**, 153
mammal bones from Wellington Caves 72, **73**
Mammals of Australia, The (1871) 56
Manly, NSW 99, 122, 135
Manta alfredi 180, **181–84**, 185
Manual of Photography, A (1857) 142
Marion's Practical Guide to Photography (1885) **126**
marlins 116, **118–19**
Marsupial Lion 148, **149**
marsupials **72**, 75, **136**, 162, 170
    see also bandicoots; Bettong; *Dasyrurus*; kangaroos; koalas; Marsupial Lion; megafauna; quolls; Tasmanian Tiger; wallabies; wombats
Mason Brothers 93
Masters, George 58, 63, 84, 161
megafauna 26, 69, 75, 173 *see also* Diprotodon
Megaptera novaeangliae **ii–iii**, v, **176–77**
Melbourne, Vic. 25, 35, 112, 180
    see also National Museum
Menura novaehollandiae **42**, 45, **168**, 169
Mesoplodon densirostris 174, **175**
Mesoplodon layardii 179
Mesoplodon sp. **4–5**
Microeca fascinans **34**, 35
micrographs 27, 125
minerals **91**
Mitchell, Thomas 69, 148
Modellimin, expedition to 49
Mola alexandrini **96–98**, 97, 99
mongooses 180, 185
Moore, Charles 24
Moree, NSW 148
Morelia spilota **46**
Moreton Bay, Qld 87
Morton, Alexander 84, 92–93
Mosaic Leatherjacket 21, **22–23**
Mosman, NSW 83
Mueller, Ferdinand von 25–26
Murray River, Vic. and NSW *see* Blandowski expedition
Murrurundi, bones at 87
Muybridge, Eadward 27
Myliobatis aquila **44**, 45

National Museum 6, 53, 64, 66, 154
natural history 21, 25–27
Natural History Museum 21, 40, 99
natural science 21, 25–26, 28
Natural Science: A Monthly Review of Scientific Progress (journal) 164
Neoceratodus forsteri 60, **61–63**, 63
Neotrigonia strangei 46, **47**
New Britain 93
New Caledonia 83
New Guinea 92–93
New South Wales Advertiser 157
New South Wales State Library 53
New York, USA 48
Noisy Friarbird nest and eggs **142**
North, Alfred 35, 79
Northern Swamp Wallaby 84, **85**
Notolabrus tetricus 21, **23**
Nototherium 75, 173
Nyeri Nyeri people 49, 53

O'Grady, Michael 59
Olliff, Sidney 78
On the Origin of Species (1859) 25–26, 78, 154
Orca **43**, 45
Orectolobus maculatus **152–53**
Orectolobus ornatus **10**, 13
Ornithorhynchus anatinus **166**
Osteological Gallery v, 101, 173, 177
Owen, Richard
    description of the Marsupial Lion 148–49
    evolution debate 55, 154
    examination of fossils 69
    exchanges with 64
    in Wellington Caves 71, 75
Oxford evolution debate 154

Pacific culture 92–93
Palmer, Edward 104
Papua New Guinea v, 93
Paris Universal Exhibition 71, 169
Parkes, Henry 180
Parkinson, Sydney 26
Pearl Perch **19**
Pectinia lactuca **128**, 129
pelicans **88–89**
Pempheris compressa **131**
Pennant Fish **18**, 19
perch **18**, **19**
Peroryctes broadbenti v, **vi**
Phascolarctos cinereus 170, **171**
pheasants **24**
Philemon corniculatus **142**
Phoenicopterus roseus **178**, 179

'Photographs of specimens in Australian Museum' 64
photography *see also* glass plates
    aesthetic style of 112
    albums **110**, 112–13, **113**, 114, **116–17**
    as image, object and information 114
    description and digitisation of 114
    field photography 28
    first studio 36
    for leisure 116, 125
    framing 136
    growing use of 122
    in Australia 126–27
    in scientific illustration 30
    introduction to the museum 6–7
    invention of 2
    lighting 136
    mise-en-scène 132
    museum photography 1857–93 32–38
    process of 2–3
    register of negatives **103**
    scientific 27–28
    Sydney Morning Herald on 121
photolithographs 30
photomicrography 27, 125
Phyllacanthus parvispinus **1**, 2
Physignathus lesueurii **46**
Pig-footed Bandicoot 64, 107
Pilumnus australis **iv**, v
pineapplefish **18**, 19
Pittard, Simon 15, 53
Platax teira **20**, **21**
Platypus **166**
Poodytes carteri **160**, 161
Port Jackson, NSW 139
Port Moresby, PNG v
'Portrait of Yarree-Yarree Aborigines' 49
prawns **139**
preparation of specimens *see* articulation; casting; staging displays; taxidermy
Priacanthus macracanthus 21, **22**
printing photographs 147, **147**
Prionocidaris australis **90**, 91
Pseudonaja textilis **86**, 87
Pseudorhombus jenynsii **18**, 19
Ptilonorhynchus violaceus **121**, 122
pufferfish **18**, 19
Pygmy Sperm Whale **54**, **94**, **134–35**
pythons 30, **30–31**, **46**

quartz **91**
Queensland 63, 114
quolls v, 162, **164–65**

Ramsay, Edward
    articulation and 179

Catalogue of Echinodermata in the Australian Museum (1885) 2
    collecting ethnographic items 92
    Curator's Report 122
    description of animals v, 6, 91, 99
    field collecting 66, 107
    Gerard Krefft and 77
    Henry Barnes and 66
    in 1885 **77**
    instructions for taxidermy 161–62
    photographic program 7, 35–36, 45
    purchase of gorilla specimens 157
    Scott sisters and 27
    vivisection by **109**
    work on classification 103, 104
Rat-kangaroo **58**
rays v, **vii**, **44**, 45 *see also* Reef Manta Ray
Records of the Australian Museum (journal) 78, 129
Red Kangaroo **150**, 153
Reef Manta Ray 180, **181–84**, 185
religion 25–26
'Return of Indigenous Cultural Property' 173
rhinoceros **102**
Rhynochetos jubatus **79**, 83
robins **34**, 35
Rockingham Bay, Qld 77
Rocks, The, NSW 12
rodents 72, **73**
Rohu, Ada 153
Roundface Batfish **20–21**
Royal Botanic Garden, Sydney 24, 25, 92
Royal Geographical Society 154
Royal Hotel, Sydney 126
Royal Society 63
Royal Zoological Society of New South Wales 173, 179

Sadlier, Ross 40
Sagmariasus verreauxi **138–39**
Sarcophilus harrisii **136**
Satin Bowerbird **121**, 122
Saville-Kent, William 129
Schaller, George 157
Schoenicola platyura **160**, 161
science *see* natural science
scientific photography 27–28
Scleropages leichardti **66**
Scott, Harriet and Helena 27, 30, 83
sculptures **92**
Sea Fan **120**, 122
seaperch **18**, 19
sea stars **110**, **113**
sea urchins **1**, 2, **90**, 91

secularisation of science 25–26
sharks 7, **8–9**, **104–105**, **140** *see also* wobbegongs
Shea, Glenn 40
Short-beaked Echidna 158, **159**
shovel-nosed crays **138**, 139
siderite **91**
Simoselaps littoralis 46, **47**
skeletons *see* articulation; Osteological Gallery
skinks **86**, 87
Slate Pencil Urchin **1**, 2
Smallscale Bullseye **131**
Smalltooth Flounder **18**, 19
Smith, John 16
snakes **78**, 83–84, **86**, 87, 180 *see also* pythons; West Coast Banded Snake
Snakes of Australia (1869) 30, **30**, 31
snapper **11**, 13
snowflakes 125
Solomon Islands, cultural objects from 92
Solomon Islands Skink **86**, 87
Southern Cassowary **76**, 77
Southern Hairy-nosed Wombat 28, **29**
Southern Saratoga **66**
Southern Shovel-nosed Cray **138**, 139
Spalding, Edward 174
spectroscopy 167
Sperm Whale **14–15**, 174 *see also* Pygmy Sperm Whale
Spinifex Bird **160**, 161
Spotted Bigeye 21, **22**
Spotted Wobbegong **152–53**
staging displays
    at International Exhibitions 66, 169–70
    for photographs 132
    of gorillas 154, 157
Stars-and-stripes Pufferfish **18**, 19
State Library of New South Wales 53
stick insects **viii**, 2
stingrays *see* rays
Stout Longtom 21, **23**
Strange's Trigonia 46, **47**
Strap-toothed Beaked Whale 179
Striate Anglerfish **19**
Striated Fieldwren **151**, 153
Striped Marlin 116, **118–19**
Sumatran Rhinoceros **102**
sunfish **96–98**, 97, 99
Superb Lyrebird **42**, 45, **168**, 169
Swainson, William **107**, 162
Swamp Wallaby **58**

Sydney, NSW
    Alexander Macleay in 12
    Bridge Street 126
    Broken Bay 139
    Clontarf Beach 180
    Intercolonial Exhibition 64
    International Exhibition 66, 92
    Royal Botanic Garden 24, 92
    Royal Hotel 126
    sea urchin found near 91
Sydney Harbour 139, 174
Sydney Mail 157, 185
Sydney Morning Herald 55, 63, 83–84, 116, 148

Tachyglossus aculeatus 158, **159**
Talbot, Henry Fox 27, 125
Tallfin Flyingfish 21, **22**
Taronga Zoo 173, 179
Tasmanian Devil **136**, 162
Tasmanian Tiger **52**, 53, 161–62, 164, 180
taxidermy *see also* articulation; casting; staging displays
    arsenic in 166–67
    field treatment for 161
    Henry Barnes' work in 6, 64, 66–67, 158
    John A Thorpe's work in 66, 99, 158
    learning 153
    of gorillas 154, 157
    process of 151, 158, 161–62, 166–67
    use in nature photography 170
    women working in 153
    workshop 122, 158
taxonomy *see* classification
theology 25–26
Thomson, Alexander 71
Thopha saccata 142, **143**
Thorpe, John A **45**, 66, 99, 158
Thylacinus cynocephalus (Thylacine) **52**, 53, 161–62, 164, 180
Thylacoleo carnifex 148, **149**, 173
Tiger Shark **140**
Tiliqua scincoides **46**
tintypes 130
Tomistoma krefftii **39**, 40
Tost, Charles 71, 153, 185
Tost, Jane 153
Transformations (1864, 1898) 27
tree-kangaroos **6**, 114, **115**
Tripletail **19**
Trygonorrhina fasciata v, **vii**
turtles **106**, 107, **108**
Tweed River, NSW 79
Tylosurus gavialoides 21, **23**

'type' specimens v, 39, 103, 135, 153
Typhlopidae family **78**, 83

United States 147
urchins **1**, 2, **90**, 91

Vanuatu 129
Varanus salvator **172**, 173
Vermicella sp. **78**, 83
Victoria 6, 28
Visitors' Guide to Sydney (1872) 56
vivisection **109**
'V negative' collection 111
Vombatus ursinus 56, **57**, **82**, 83
von Mueller, Ferdinand 25–26

Waite, Edgar Ravenswood v, 125
Wall, Thomas 77
Wall, William Sheridan 15, 32, 174
Wallabia bicolor **58**
Wallabia bicolor mastersii 84, **85**
wallabies **58**, 84, **85**
Ward, Henry 170
water dragons **46**
Weekly Times, Hobart 127
Wellington Caves, NSW
    Henry Barnes' control of excavation 66
    Henry Barnes' personal photographs 116
    impact of 26, 33, 55
    landscape **70–71**, **127**
    remains in **68**, 69, 71, **72–74**, 75, 148
Wentworth, William Charles 67
West Coast Banded Snake 46, **47**
Western Gorilla **155**
wet plate process 130, 141, **141**
whales **132**, 144, **145–46**, 179–80 *see also* beaked whales; Humpback Whale; Orca; Sperm Whale
Whitelegge, Thomas 122
White Shark 7, **8–9**, **104–105**
'Wild Planet' Gallery 174
Wilson, Glen 32
wobbegongs **10**, 13, **152–53**
wombats 28, **29**, **74**, 75 *see also* Common Wombat
Wood, William 39
wrasse 21, **23**
wrens **151**, 153

Zoological and Acclimatisation Society of Victoria 102
Zoological Society of London 40, 53, 63
Zoological Society of New South Wales 173, 179

A NewSouth book
Published by
NewSouth Publishing
University of New South Wales Press Ltd
University of New South Wales
Sydney NSW 2052
AUSTRALIA

newsouthpublishing.com

Published in association with the Australian Museum in conjunction
with the exhibition *Capturing Nature: Early scientific photography
1857–1893*.

www.australianmuseum.net.au

© Australian Museum Trust 2019
First published 2019

10 9 8 7 6 5 4 3 2 1

 A catalogue record for this
book is available from the
National Library of Australia

ISBN 9781742236209 (paperback)

Design Pfisterer + Freeman
Front cover image Southern Cassowary, *Casuarius casuarius*.
Photographed at the Australian Museum by Henry Barnes.
Inside front cover images Banded Wobbegong, *Orectolobus
ornatus,* and Northern Swamp Wallaby, *Wallabia bicolor mastersii*.
Photographed at the Australian Museum by Henry Barnes.
Inside back cover images Crabs, *Pilumnus australis*, and nest and
eggs of the Satin Bowerbird, *Ptilonorhynchus violaceus*. Photographed
at the Australian Museum by Henry Barnes, jnr.
Back cover images Clockwise from top left: Australian Freshwater
Crocodile, *Crocodylus johnstoni*; Short-beaked Echidna, *Tachyglossus
aculeatus*; Common Wombat, *Vombatus ursinus*; Eastern Fiddler
Ray, *Trygonorrhina fasciata*; Giant Bandicoot, *Peroryctes broadbenti*.
Photographed at the Australian Museum by Henry Barnes.
Printer 1010 Printing International Limited, China

All reasonable efforts were taken to obtain permission to use copyright
material reproduced in this book, but in some cases copyright could
not be traced. The author welcomes information in this regard.

This book is printed on paper using fibre supplied from
plantation or sustainably managed forests.